Blade Element Rotor Theory

Blade Element Rotor Theory presents an extension of the conventional blade element rotor theory to describe the dynamic properties of helicopter rotors. The presented theory focuses on accurate mathematical determination of the forces and moments by which a rotor affects its rotorcraft at specified flight conditions and control positions. Analyzing the impact of a blade's non-uniform properties, the book covers blade twisting, the non-rectangular planform shape of a blade, and inhomogeneous airfoil along the blade. It discusses inhomogeneous induced airflow around a rotor disc in terms of the blade element rotor theory. This book also considers the impact of flapping hinge offset on the rotor's dynamic properties.

Features

- Focuses on a comprehensive description and accurate determination of the rotor's aerodynamic properties

- Presents precise helicopter rotor properties with inhomogeneous aerodynamic properties of rotor blades

- Considers inhomogeneous distribution of induced flow

- Discusses a mathematical model of a main helicopter rotor for a helicopter flight simulator

This book is intended for graduate students and researchers studying rotor dynamics and helicopter flight dynamics.

Blade Element Rotor Theory

Pylyp Volodin

MARKET-MATS ltd.

CRC Press
Taylor & Francis Group
Boca Raton London New York

CRC Press is an imprint of the
Taylor & Francis Group, an **informa** business

First edition published 2023
by CRC Press
6000 Broken Sound Parkway NW, Suite 300, Boca Raton, FL 33487-2742

and by CRC Press
4 Park Square, Milton Park, Abingdon, Oxon, OX14 4RN

CRC Press is an imprint of Taylor & Francis Group, LLC

ISBN: 978-1-032-28303-6 (hbk)
ISBN: 978-1-032-28304-3 (pbk)
ISBN: 978-1-003-29623-2 (ebk)

DOI: 10.1201/9781003296232

Typeset in Nimbus font
by KnowledgeWorks Global Ltd.

Publisher's note: This book has been prepared from camera-ready copy provided by the authors.

Contents

Preface

Different approaches and theories have been developed to describe properties of a helicopter rotor demanded by the impetuous development of rotary-wing aircraft. However, none of them is self-sufficient to provide comprehensive enough description of rotor operation in terms of interaction with a rotorcraft. When we have been challenged with creation of a simulator for a single rotor helicopter, the combination of the approaches was sought for a rotor model that can be realized by real-time computation. The blade element rotor theory established during the last century was chosen as predominant because it provides an explanation of the swashplate mechanism operation, which is required to simulate the rotor control. The results of the enhancement of this theory were found interesting to present them in open sources. Despite the fact that the book originated from simulation purposes, it does not discuss the simulation algorithms, but focuses on the explanations of helicopter rotor operation as a main helicopter element that enables rotorcraft flight.

The book's subject matter evolves and extends the blade element rotor theory to provide a more precise mathematical description of the rotor properties. The main idea behind the theory is that each part of a rotor blade is considered a wing, which is able to create a lift force and an aerodynamic drag during rotation around the rotor axis. Total lift force of all elements of all blades transfers on the rotorcraft enabling its flight in the airspace. However, the free flapping blade construction accompanied by a swashplate mechanism makes the rotor operation much more complex and cannot be simplified only to such interpretation. The blade lift forces provoke the blades to flap around the flapping hinges; the blade flapping motion changes the action of the total aerodynamic force depending on current flight conditions and swashplate position. The blade flapping as well depends on blade aerodynamic properties, which are usually not identical along the blade spar. The offset of flapping hinges, where the blades are attached to the rotor, contributes to the complicity of rotor performance by creating the force moments acting on the rotorcraft. The swashplate mechanism, which is intended to control the rotor with further control over the rotorcraft, more complexly affects the rotor due to the mentioned aspects; thus it must be analyzed precisely at a whole range of the flight conditions.

These challenges are discussed in the book to find a comprehensive rotor mathematical model which accurately describes the rotor performance in the whole range of flight conditions. The blade flapping during rotation around the rotor axis is considered depending on swashplate position, flight conditions, and blade properties. The impact of the swashplate on the blade flapping motion is analyzed to explain the control over the rotor performance by the swashplate. There it is discussed how the aerodynamic forces, which are created by the rotated blades, are transferred on the rotorcraft via rotor construction elements. A general destination of the issue is the determination of forces and moments of the rotor acting on its rotorcraft at different flight conditions and swashplate positions. This book enables readers with basic knowledge of aerodynamics and rigid body dynamics to comprehensively understand the helicopter rotor operation and how the rotor interacts with its rotorcraft.

Symbols

a_0	blade cone angle
a_1	longitudinal blade cone tilt
$\vec{A}^{(h)}$	translational acceleration of rotor system with coordinates in blade-hinge frame
b_1	sideways blade cone tilt
$B^{(n)}$	blade characteristic parameter of power n
$B_m^{(n)}$	blade-hinge characteristic parameter of power n and m
C_{De}	element profile drag coefficient
$C_D(y_b)$	function of element profile drag coefficient on element position
$C_D^{(n)}$	characteristic blade profile drag coefficient of power n
c_e	element chord length
$c(y_b)$	function of element chord length on element position
C_{Le}	element lift coefficient
C_{Le}^{α}	element lift coefficient slope
$C_L^{\alpha}(y_b)$	function of element lift force coefficient slop on element position
C_H	longitudinal force coefficient
C_{mx}	lateral hub moment coefficient
C_{my}	longitudinal hub moment coefficient
C_{mz}	resistance moment coefficient
C_P	rotor power coefficient
C_Q	rotor torque coefficient
C_S	side force coefficient
C_T	rotor trust coefficient
D_0	coefficient for calculations of a_1 and b_1
D_μ	coefficient for b_1 calculation
D_k	flapping compensation impact coefficient for b_1 calculation
D_{k2}	flapping compensation impact coefficient for a_1 calculation
$D_k^{(a0)}$	flapping compensation impact coefficient for a_0 calculation
$D_{k2}^{(a0)}$	coefficient for a_0 calculation with k^2
dy_b	blade element length
dD	blade element profile drag force
$d\vec{F}_e^{(inter)}$	internal force acting on blade element
dI	tensor of element moment of inertia in blade-hinge frame
$dI^{(b)}$	tensor of element moment of inertia in blade fixed frame
dI_C	tensor for Coriolis moments in blade-hinge frame
$dI_C^{(b)}$	tensor for Coriolis moments in blade fixed frame
$dI_{xx}^{(b)}$	blade element moment of inertia around blade transverse x_b-axis
$dI_{yy}^{(b)}$	blade element moment of inertia around blade longitudinal y_b-axis

$dI_{zz}^{(b)}$	blade element moment of inertia around blade normal z_b-axis
$dI_{Cxx}^{(b)}$	element of tensor for Coriolis moments around blade transverse x_b-axis
$dI_{Cyy}^{(b)}$	element of tensor for Coriolis moments around blade longitudinal y_b-axis
$dI_{Czz}^{(b)}$	element of tensor for Coriolis moments around blade normal z_b-axis
dL	blade element lift force
dm	blade element mass
dM_{AD}	blade element pitching moment
$d\vec{M}_e^{(exter)}$	superposition of external force moments acting on element
$d\vec{M}_e^{(inter)}$	internal torque caused by internal force acting on element
$d\vec{R}$	blade element aerodynamic force
dS	blade element planform area
dX	tangential component of blade element aerodynamic force (along x-axis)
dY	longitudinal component of blade element aerodynamic force (along y-axis)
dZ	normal component of blade element aerodynamic force (along z-axis)
$\vec{e}_x$	unit vector of tangential x-axis of blade-hinge frame
$\vec{e}_y$	unit vector of longitudinal y-axis blade-hinge frame
$\vec{e}_z$	unit vector of normal z-axis blade-hinge frame
$\vec{F}_{dA}$	aerodynamic force per elementary area
$\vec{F}_{FH}$	blade force on flapping hinge
$\vec{F}_{j,k}^{(inter)}$	internal force of interaction between two blade particles
$\vec{F}_{i\Sigma}^{(exter)}$	superposition of external forces acting on i-th element
$\vec{F}_{react}(y_b)$	sum of reaction forces acting on an element at y_b position
H	rotor longitudinal force
$\vec{g}^{(h)}$	gravitational acceleration vector with coordinates in blade-hinge frame
I	blade moment of inertia tensor in blade-hinge frame
I_C	tensor for blade Coriolis moments
$I_{Cxx}^{(b)}$	element of tensor for Coriolis moments around blade transverse x_b-axis
$I_{Cyy}^{(b)}$	element of tensor for Coriolis moments around blade longitudinal y_b-axis
$I_{Czz}^{(b)}$	element of tensor for Coriolis moments around blade normal z_b-axis
$I_{xx}^{(b)}$	blade moment of inertia around blade transverse x_b-axis
$I_{xx}^{(r)}$	blade moment of inertia axis parallel to x_b-axis and passing hub center
$I_{yy}^{(b)}$	blade moment of inertia around blade longitudinal y_b-axis
$I_{zz}^{(b)}$	blade moment of inertia around blade normal z_b-axis
$I_{zz}^{(r)}$	blade moment of inertia around rotor axis
J	blade mass moment around flapping hinge
k	flapping compensation coefficient
k_e	effective flapping compensation coefficient
$k^{l_{FH}}$	sensitivity of effective flapping compensation coefficient to flapping hinge offset
L_B	blade length
l_{FH}	blade flapping hinge offset

$\vec{l}_{FH}^{(h)}$	flapping hinge center position in blade-hinge frame
$\vec{l}_{FH}^{(r)}$	flapping hinge center position in rotor frame
m_B	blade mass
$M^{(r \to h)}$	transformation matrix from rotor frame to blade-hinge frame
$M^{(b \to h)}$	transformation matrix from blade fixed frame to blade-hinge frame
$m_{ad}^{(n)}$	characteristic pitching moment of blade of power n
M_{FH}	flapping hinge reaction moment with coordinates in blade-hinge frame
M_{HUBx}	lateral hub moment (around x_r-axis)
M_{HUBy}	longitudinal hub moment (around y_r-axis)
M_{HUBz}	resistance moment (around z_r-axis)
M_{onFH}	blade moment on flapping hinge
n	number of blades of rotor system
N_p	number of point particles in blade
n_p	number of point particles in blade element
O_r	rotor hub center
O	origin of blade fixed and blade-hinge frames, flapping hinge center
$\vec{p}^{(b)}$	position of blade point in blade fixed frame
$\vec{p}$	position of blade point in blade-hinge frame
$\vec{p}_{ac}^{(b)}$	position of blade element aerodynamic center in blade fixed frame
$\vec{p}_e$	position of element center of mass in blade fixed frame
$\vec{p}_j^{(b)}$	position of j-th point particle in blade fixed frame
$\vec{p}_j$	position of j-th point particle in blade-hinge frame
$\vec{p}_{lug}^{(i)}$	position of i-th lug in blade-hinge frame
$P_{consume}$	rotor system consumption power
r	section radius of blade element
$\vec{r}$	radius-vector of blade element position in rotor frame
R	rotor disc radius
$\vec{R}$	total rotor force
$\vec{R}_{FH}^{(i)}$	flapping hinge reaction force on i-th lug in blade-hinge frame
$\vec{R}_{pitch}$	blade pitch force with coordinates in blade-hinge frame
R_{pitch}	normal coordinate of blade pitch force (along z-axis)
R_{slip}	slip stopper reaction force (along x-axis)
$\vec{R}_{SP}$	blade force on swashplate
R_{Yaxle}	longitudinal flapping hinge axle reaction force (along y-axis)
R_{Zaxle}	normal flapping hinge axle reaction force (along z-axis)
S	rotor side force
S_B	blade planform area
$slip_e$	element sideslip angle
T	rotor thrust
Q	rotor torque (around z_r-axis)
$\vec{v}$	absolute velocity of blade point with coordinates in blade-hinge frame
$\vec{V}$	hub velocity center with coordinates in rotor frame
V_e	element airspeed

$\vec{V}_e$	element air velocity with coordinated in blade-hinge frame
V_{ex}	tangential coordinate of element air velocity (along x-axis)
$\bar{V}_{ex}$	normalized tangential coordinate of element air velocity to ωR
V_{ey}	longitudinal coordinate of element air velocity (along y-axis)
$\bar{V}_{ey}$	normalized longitudinal coordinate of element air velocity to ωR
V_{ez}	normal coordinate of element air velocity (along z-axis)
$\bar{V}_{ez}$	normalized normal coordinate of element air velocity to ωR
$V_i(y_b, \psi)$	induced velocity of blade element in y_b, ψ position parallel to z_r-axis
$\vec{v}_j$	absolute velocity of j-th point particle with coordinates in blade-hinge frame
V_x	speed of forward motion
V_y	velocity of vertical motion
x	tangential axis of blade-hinge frame
$x_{ac}^{(b)}$	transverse coordinate of element aerodynamic center
$\bar{x}_{ac}^{(n)}$	characteristic normalized position of blade aerodynamic center of power n
x_b	blade transverse axis of blade fixed frame
$x_{lug}^{(i)}$	tangential position of i-th lug (along x-axis)
$x_{pitch}^{(b)}$	transverse position of pitch horn joint with pitch link in blade fixed frame
x_r	rotor longitudinal axis of rotor frame
y	longitudinal axis of blade-hinge frame (coincides with y_b-axis)
y_b	blade longitudinal axis of blade fixed frame
$y_{pitch}^{(b)}$	longitudinal position of pitch horn joint with pitch link in blade fixed frame
y_r	rotor lateral axis of rotor frame
z	normal axis of blade-hinge frame
z_b	blade normal axis of blade fixed frame
z_r	rotor axis
α_e	element attack angle
β	blade flapping angle
γ	parameter equals $\rho S_B \bar{C}_L^\alpha R^3 / (2 I_{zz}^{(r)})$
ϵ	ratio of moments of inertia $I_{xx}^{(b)} / I_{zz}^{(r)}$
ϵ_I	sensitivity of ϵ to $\bar{l}_{FH}$
ϵ_μ	coefficient-multiplier for μ equals $B_1^{(2)} / B_2^{(4)}$
ε_e	angle between an element air velocity projection on reference plane and x-axis
$\varepsilon_{\Omega x}$	rate of Ω_x change $(d\Omega_x / dt)$
$\varepsilon_{\Omega y}$	rate of Ω_y change $(d\Omega_y / dt)$
$\vec{\varepsilon}_\omega^{(h)}$	rotor shaft angular acceleration with coordinates in blade-hinge frame $M^{(r \to h)}(d\vec{\omega}/dt)$
$\vec{\varepsilon}_\Omega^{(h)}$	rotor system angular acceleration with coordinates in blade-hinge frame $M^{(r \to h)}(d\vec{\Omega}/dt)$
$\bar{\varepsilon}_{\Omega x}$	normalized angular accelerations of rotor system around x_r-axis $(\varepsilon_{\Omega x}/\omega^2)$
$\bar{\varepsilon}_{\Omega y}$	normalized angular accelerations of rotor system around y_r-axis $(\varepsilon_{\Omega y}/\omega^2)$

θ_0 collective pitch angle

θ_1 lateral cyclic pitch angle

θ_2 longitudinal cyclic pitch angle

$\lambda(y_b, \psi)$ inflow ratio of element at position y_b, ψ

$\lambda^{(n)}(\psi)$ basic characteristic inflow ratio of power n in blade azimuthal position ψ

$\lambda^{(n,m)}(\psi)$ characteristic inflow ratio of power n and m in blade azimuthal position ψ

$\lambda_0^{(n,m)}$ average $\lambda^{(n,m)}(\psi)$ around rotor disc

$\lambda_0^{(n)}$ equals to $\lambda_0^{(n,0)}$

$\lambda_0^{tw(n)}$ average characteristic inflow ratio of power n associated with blade twisting

$\lambda_{ak}^{(n,m)}$ cosine Fourier coefficient of k-th harmonic of $\lambda^{(n,m)}(\psi)$ inhomogeneity over disc

$\lambda_{ak}^{(n)}$ equals to $\lambda_{ak}^{(n,0)}$

$\lambda_{ak}^{tw(n)}$ cosine Fourier coefficient of k-th harmonic of inflow ratio associated with blade twisting

$\lambda_{bk}^{(n,m)}$ sine Fourier coefficient of k-th harmonic of $\lambda^{(n,m)}(\psi)$ inhomogeneity over disc

$\lambda_{bk}^{(n)}$ equals to $\lambda_{bk}^{(n,0)}$

$\lambda_{bk}^{tw(n)}$ sine Fourier coefficient of k-th harmonic of inflow ratio associated with blade twisting

μ advance ratio

ρ volumetric mass density of surrounded airspace

σ rotor solidity

φ blade pitch (incidence) angle

φ_e element pitch (incidence) angle

$\varphi_{tw}(y_b)$ a blade twisting (function)

$\varphi^{(n)}$ basic characteristic twisting angle of power n

$\varphi_m^{(n)}$ characteristic twisting angle of power n and m

ψ azimuthal angle of direction or blade position $[0, 2\pi)$

ω speed of rotor shaft rotation

$\vec{\omega}$ angular velocity of shaft rotation with coordinates in rotor frame

$\vec{\omega}_h$ angular velocity of blade-hinge frame relative to inertial coordinate system

$\vec{\Omega}$ angular velocity of rotor system rotation with coordinates in rotor frame

Ω_x angular velocity of lateral rotation of rotor system around x_r-axis

$\bar{\Omega}_x$ normalized angular velocity of rotor system around x_r-axis (Ω_x/ω)

Ω_y angular velocity of longitudinal rotation of rotor system around z_r-axis

$\bar{\Omega}_y$ normalized angular velocity of rotor system around y_r-axis (Ω_y/ω)

Acknowledgments

I wish to thank the Chief of Department No. 120 of MARKET-MATS Ltd., Mr. Stanislav Krol, for support of this work and for permission to publish the results, which are presented here. I thank my fellow programming-engineers of Department No. 120 of MARKET-MATS Ltd. for the teamwork in creation of software for flight simulators. I gratefully thank Mr. Igor Yatsyshyn for the thorough adjustment and complex debugging of the flight simulators and for productive mutual work. I especially thank helicopter pilot Col. Vladimir Pastukhov, for comprehensive discussions of the helicopter flight dynamics and for thorough testing of helicopter flight simulator.

1 Introduction

Flight simulators for training of aircraft pilots are useful and suitable equipment to train pilot cadets as well as to maintain and enhance the skills of operating pilots. Flight simulators enable to achieve the piloting skills before real flight performance and to train the operations in case of onboard equipment failures and in an emergency situation, which are difficult to perform in a real flight. They enable to retain the flight resources of a correspondent aircraft and to reduce fuel spends. The important requirement of flight simulators is to provide the adequate flight simulation of a correspondent aircraft according to the position of the aircraft controls and the simulated flight conditions.

An essential part of the simulator software is a flight dynamics application, which performs computations of an aircraft motion in the airspace. The flight dynamics generally solves the equations of motion of the aircraft as a rigid body under the actions of forces and moments, which appear at the interaction of the aircraft with the air. These forces and moments are supported by power of an engine or engines of an aircraft. The flight dynamics of an aircraft can be realized as a set of elements, each of which represents a separate source of forces and moments affecting the simulated aircraft. For example, the flight dynamics of a helicopter with a single-rotor configuration can be simplistically decomposed into the following separate force source elements: a main rotor, a tail rotor, a fuselage, gears as interaction with the ground, and the gravity. Each element is described by a correspondent model, which computes element forces and moments acting on the aircraft and is based on the nature of this element. Each element model is a part of the aircraft flight dynamics software. All element models compose the superposition of forces and moments, based on which the aircraft motion is computed.

Each element model represents a set of algorithms and computations, which form output data, which affect the aircraft flight dynamics, based on input data, which represent current flight conditions, position of aircraft controls, and states of other models. The output data usually represents the forces and moments caused by the element and states required by other elements. An element model operates with configuration parameters, values of which remain constant and describe a specific instance of an element used in flight dynamics of a specific aircraft.

The challenge of a flight dynamics developer of a specific aircraft is to find appropriate models and configuration parameters of elements, which are required for the flight simulation of the aircraft with stated accuracy. The flight simulator with its flight dynamics software is qualified by the accuracy of reproducibility of the aircraft flight performances as well as by the validation performed by expert pilots.

An algorithm of a model of a flight dynamics element, which can be performed as a sequential set of mathematical calculations, is preferable for a developer. Such approach enables analytical determination of values of the model configuration parameters based on given flight performances of a simulated aircraft. The fast

DOI: 10.1201/9781003296232-1

determination of the configuration parameter simplifies multiple adjustments of the whole simulator during adjusting and complex debugging. Multiple changes of the configuration parameters may implicitly affect some other flight performances of the simulator, which can remain hidden for current attention: it is possible with this approach to predict what features would be affected by an introduced change. However, nature of some flight dynamics elements might be too complicated to realize in such a way despite the advantages.

Models of complicated elements, which cannot be realized with mathematical sequential calculations, can use iterative algorithms in order to generate appropriate output data. Despite the ability to perform calculations of complicated models, it is very hard to operate with such methods in adjustments of the model as a part of the whole flight dynamics application. A developer must be aware of convergence and the accuracy of the algorithms; the developer must care about computational resources. It is hard to find the appropriate values of the model configuration parameters based on expected outputs; an exhaustive search might be complicated due to the large set of the configuration parameters and might give a physically senseless result. It is especially complicated if other flight dynamic elements also use iterative algorithms. It is highly possible that any change has implicitly hidden impacts on some other flight performances, which are out of current attention.

The developer searches the tradeoff between these two approaches to find appropriate, accurate, and adequate models. In any case, the developer must understand the dynamic and aerodynamic background of each element and the whole flight dynamics of the simulated aircraft.

A comprehensive and adequate model of a main helicopter rotor is the essential part of flight dynamics software, which is used in a simulator of a helicopter with a single-rotor configuration. The rotor model, together with models of other elements and systems of such a helicopter, must provide simulation of control over the helicopter, which is as close as possible to the real control: this task is a great challenge of the rotor model developer.

The different approaches and theories were developed to describe dynamic properties of helicopter rotors during the last century demanded by the impetuous development of rotary-wing aircraft. There were momentum theory, blade element rotor theory, and vortex theories. However, none of these theories is self-sufficient as a base for a model of a rotor with a swashplate mechanism, which comprehensively and accurately computes forces and moments affecting the rotorcraft depending on the control positions and flight conditions. Adequate helicopter control, which is performed via a swashplate mechanism of a main rotor, must be emphasized for a flight simulator of such a helicopter.

Hereafter are presented principles of dynamics and aerodynamics of a helicopter rotor, which were developed as the base of a rotor model for a flight simulator of a Mi-24 helicopter. These principles are based on the conventional blade element rotor theory, which had unexplored potential extensions. The main destination of the research was to find suitable approaches, which can provide the comprehensive and accurate simulation of a helicopter rotor at a whole range of flight conditions. This conventional theory was evolved to describe a wide range of rotor configurations, such as blade twisting, blade non-uniform properties, inhomogeneous induced flow. Despite the presented issues are originated from the simulation purposes, the simulation algorithms are not discussed here, but the comprehensive description and accurate determination of the rotor aerodynamic properties are in focus within the issue.

Dynamically steady states of a rotor appear quite seldom in real simulation. The transient processes and control system time delay are very important for correct modeling of rotor response at the rotorcraft control simulation. However, these transient processes remain out of the current discussion; only the rotor steady states are considered here as a basis of the rotor description. The results of the rotor steady states were found quite interesting to present them in open sources.

2 General Statements

A rotor system is defined here as the combination of rotary wings (blades) that generate the aerodynamic lift forces in order to: support the weight of an aircraft in a flight; counteract the aerodynamic drag of the aircraft at forward flight; and control over the aircraft motion. A rotorcraft is defined here as a heavier-than-air aircraft which uses such a rotor system to perform a flight in an airspace. Generally, a rotor system transforms energy of a rotorcraft engine into energy required for a rotorcraft flight in an airspace. It is assumed hereafter that a rotor system has defined upper and below sides; an engine of a rotorcraft is usually located below the rotor system.

The rotor system, which properties will be studied further, is described in this chapter. Here are specified approaches that are used for these studies.

2.1 CONSTRUCTION AND COMPONENTS

A helicopter rotor system is discussed here with the following construction and components.

2.1.1 The rotor system consists of the following main components: a rotated rotor shaft; a rotor hub, which is fixed to the rotor shaft; rotor blades, which are attached to the rotor hub; and a pitch control mechanism of the rotor blades (fig. 2.1).

2.1.2 The rotor shaft is permanently rotated around the rotation axis (or the rotor axis) that is directed along the shaft axis. The permanent rotation of the shaft is kept by applying a required torque, which is generated by a certain engine. The shaft rotates counter-clockwise direction viewing from the upper side of the rotor system. The magnitude of the angular velocity of the shaft (the shaft angular speed) is denoted by ω, which takes positive values.

2.1.3 The rotor system has more than one blade; a number of the rotor blades is denoted n ($n > 1$). The blades are identical, which means that all blades have similar physical and aerodynamic properties: geometry; mass; moments of inertia; and aerodynamic airfoil. The identity of the blades ensures that each blade performs identical motion under identical conditions. The blades are considered here as absolutely rigid bodies with straight spars. A blade length is a distance from the blade edge on the root side to the blade tip and is denoted by L_B. A blade is considered as a rotary-wing with a high wing aspect ratio; this means that the length of the blade is much longer than its chord as well as its thickness.

2.1.4 Each blade is attached from its root side to the rotor hub via a flapping hinge, around which the blade is free to turn (fig. 2.1). The flapping hinges are situated on the rotor hub at equal distances to the rotor axis in axisymmetric order around the rotor axis and lay in one plane, which is perpendicular to the rotor axis. This plane is called the rotor plane hereafter. The distance from the rotor axis to a flapping hinge axis is called the flapping hinge offset and is denoted l_{FH}.

DOI: 10.1201/9781003296232-2

2.1.5   If the rotating rotor blades would lay in the rotor plane, then the blades would circumscribe a disc, which is called the rotor disc hereafter. The rotor disc lays in the rotor plane. The radius of this disc is called the rotor disk radius, is denoted by R, and equals the sum of the rotor blade length and the flapping hinge offset: $R = l_{FH} + L_B$.

2.1.6   It is assumed here that an axis of each flapping hinge, around which an attached blade turns, lays in the rotor plane. The blades are attached to the hub via flapping hinges in such a way that blade spars are radially aligned to the rotor axis if the blades lay in the rotor plane and each blade spar is perpendicular to the axis of the blade flapping hinge.

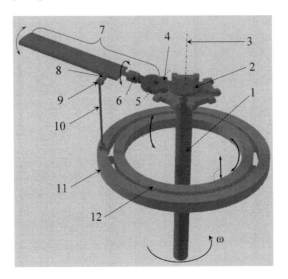

Figure 2.1   Scheme of rotor system: 1 - rotor shaft; 2 - rotor hub; 3 - rotor axis; 4 - flapping hinge; 5 - flapping hinge axis; 6 - feathering hinge; 7 - blade; 8 - blade pitch horn; 9 - swivel joint; 10 - pitch link; 11 - rotating swashplate; 12 - stationary swashplate.

2.1.7   The rotor hub is rotated together with the rotor shaft around the rotor axis and rotates the flapping hinges with attached blades. During this rotation, each blade is free to turn without friction around the axis of its flapping hinge under forces acting on the blade such as gravity, aerodynamic forces, Coriolis forces, and centrifugal forces. The motion of the blade around its flapping hinge is called the blade flapping motion.

2.1.8   Aerodynamic forces are generated along the blade spar in oncoming airflow. The aerodynamic forces of the blade depend on the blade pitch (incidence) angle, which is defined here as the angle on which this blade should be turned around its spar in such a way, that the chord of the root blade element would be parallel to the rotor plane. The blade pitch angle is denoted by φ. Blade pitch angle equals zero if the chord of the root blade element is parallel to the rotor plane; blade pitch angle is positive if the front edge of the blade root element is directed upward of the rotor system.

2.1.9   A feathering hinge, which is suited on a blade spar after the blade flapping hinge, enables the blade to turn around its spar without friction. The feathering hinge provides the possibility to change the blade pitch angle.

2.1.10   The pitch control mechanism of a swashplate (or swashplate mechanism) is used to control over pitch angles of the rotor blades during their rotation around the rotor axis. The swashplate consists of two parts: a stationary swashplate, which is mounted on the rotor pylon and a rotating swashplate, which is rotated together with the rotor shaft and the rotor blades (fig. 2.1). The rotating swashplate is connected to the stationary swashplate via a bearing. The stationary and rotating swashplates have a same plane and a same center; this center lays on the rotor axis below the rotor hub. The swashplate center can change position along the rotation axis. The swashplate plane can be arbitrarily oriented to the rotor axis in a usual range of ten degrees from perpendicular orientation. Such swashplate orientation is possible with a cardan joint of the stationary swashplate. The rotating swashplate is connected to each blade via a pitch link, which is swivel joined to a blade pitch horn. The pitch link is aligned parallel to the rotor axis. The pitch horn is rigidly fixed to its blade after the feathering hinge, is parallel to the chord of the blade root element, and posits the blade pitch angle.

2.1.11   It is assumed here that no lagging hinge is applied to any blade.

2.1.12   A center of the rotor hub is defined as an intersection point of the rotor plane with the rotor axis. This point is denoted by O_r.

2.1.13   The whole rotor system is attached to a rotorcraft. The rotor system moves together with the rotorcraft in the airspace. It is assumed here that: the airspace is undisturbed without taking into account the perturbations created by the rotor system itself; the airspace is stationary relative to an inertial coordinate system; the rotorcraft without the rotor system is a rigid body.

2.1.14   The motion of the rotor system in the airspace is described by a vector of the velocity of the rotor hub center $\vec{V}$ and a vector of angular velocity $\vec{\Omega}$ of the rotor system rotation, which the rotor system performed together with its rotorcraft. The angular velocity of the rotor shaft ω, the velocity of the hub center $\vec{V}$, the angular velocity of the rotor system $\vec{\Omega}$, and the density of the surrounded airspace ρ define the conditions of the rotor system.

2.2   COORDINATE SYSTEMS AND ORIENTATION ANGLES

2.2.1   A rotor (intermediate) frame is used to describe the dynamic properties of the rotor system in terms of rotated blades (fig. 2.2). The origin of this frame is located at the hub center (the point O_r); the rotor longitudinal x_r-axis is directed along a projection of the hub center velocity vector on the rotor plane (V_x); the rotor lateral y_r-axis lays in the rotor plane and is directed right-side viewing upright along the rotor longitudinal x_r-axis; the z_r-axis completes the right-hand coordinate system and is directed downward of the rotor system along the rotor axis.

2.2.2   Position and orientation of a blade in the rotor frame are determined by three angles (fig. 2.2).

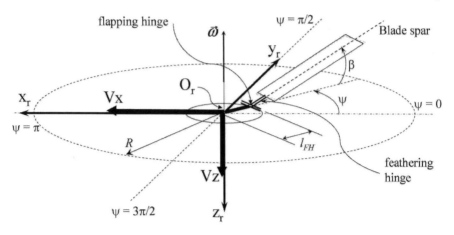

Figure 2.2   The rotor frame.

- The azimuthal angle of the blade is historically defined as the angle from the reverse direction of the x_r-axis to the blade spar projection on the rotor plane in the direction of the shaft rotation. The azimuthal angle is denoted by ψ and takes values in a range from 0 to 2π.
- The blade flapping angle is defined as the angle between the blade spar and the rotor plane and is denoted by β. The flapping angle equals zero if the blade spar lays in the rotor plane; this angle is positive if the blade is from the upper side of the rotor plane; this angle is negative if the blade is below the rotor plane.
- The blade pitch angle φ (see 2.1.8) describes how the blade is turned around the blade feathering hinge.

2.2.3   The hub center velocity is represented by coordinates $\vec{V}(V_x, 0, V_z)$ in the rotor frame: V_x is a projection of this velocity on the rotor plane and coincides with x_r-axis, therefore, the lateral y_r-coordinate equals zero; V_z is a projection of this velocity on the z_r-axis, which coincides with the rotor axis. According to the z_r-axis direction, the velocity coordinate V_z is negative if the rotor hub moves upward regarding the rotor system; V_z is positive if the rotor hub moves downward. Hereafter, the coordinate V_x is called the velocity of forward motion of the rotor system; the coordinate V_z is called the velocity of vertical motion of the rotor system.

The shaft angular velocity is represented by coordinates $\vec{\omega}(0, 0, -\omega)$ in the rotor frame. Since the axis of the shaft rotation is aligned to the z_r-axis, then the x_r- and y_r-components equal zero. The sign minus denotes the counter-clockwise shaft rotation viewing from the top side of the rotor system as stated above. Based on the definition of the blade azimuthal position, the rotor shaft angular speed can be expressed as a first derivative of the blade azimuthal position with respect to time $\omega = d\psi/dt$ with an assumption, that the projection of the hub center velocity on the rotor plane does not change its orientation relative to the inertial coordinate system.

The angular velocity of the whole rotor system is represented by coordinates $\vec{\Omega}(\Omega_x, \Omega_y, 0)$ in the rotor frame. It is assumed that the z-coordinate of this angular velocity is a part of the shaft angular velocity $\vec{\omega}$. The rotation around the y_r-axis with the velocity Ω_y is called longitudinal rotation of the rotor system; the rotation around the x_r-axis with the velocity Ω_x is called lateral rotation of the rotor system hereafter.

2.2.4 A blade fixed frame is fixed to a blade and is generally used to describe its mechanical properties (fig. 2.3). The frame origin is located at the intersection of the blade spar axis with the blade flapping hinge axis and is denoted by the point O. The y_b-axis is directed along the blade spar from the blade root toward the blade tip and is called the blade longitudinal axis. The x_b-axis is aligned parallel to the chord of the blade root element, is directed toward the blade rotational motion around the rotor shaft, and is called the blade transverse axis. The z_b-axis is perpendicular to the y_b-axis and x_b-axis, is directed downward of the rotor system to complete the right-hand coordinate system, and is called the blade normal axis.

It is supposed here that the blade longitudinal axis of the blade fixed frame is aligned to the central longitudinal principal axis of the blade inertia. Other central principal axes of the blade inertia are fixed to the blade fixed frame.

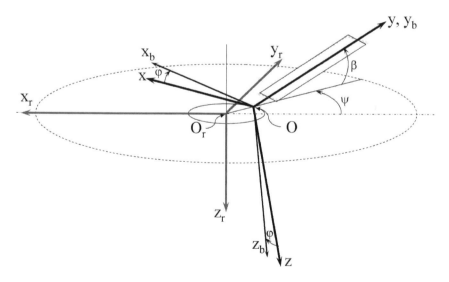

Figure 2.3 The blade fixed frame and the blade-hinge frame.

2.2.5 A blade-hinge frame is used to describe dynamic properties of a blade relative to its flapping hinge (fig. 2.3). The frame origin is located at the intersection of the blade spar axis with the blade flapping hinge axis (a point O) similar to the blade fixed frame. This intersection point is also called the blade flapping hinge center hereafter. The y-axis is directed along the blade spar from the blade root to the blade tip and coincides with the blade longitudinal y_b-axis of the blade fixed frame; therefore, it is called as well the blade longitudinal axis. However, the x-axis is aligned to the blade flapping hinge axis, lays in the rotor plane, and is directed toward the rotational motion of this flapping hinge around the rotor shaft. Because of this, the x-axis is called the tangential axis hereafter. The z-axis is perpendicular to the x- and y-axes and is directed downward of the rotor system to complete the right-hand coordinate system. In order to avoid misunderstandings with the name of blade normal axis z_b, the z-axis is called the normal axis without specification hereafter.

Directions of the axes of the blade-hinge frame are described by unit vectors: $\vec{e}_x$ is a unit vector for the tangential axis; $\vec{e}_y$ is a unit vector for the longitudinal axis; $\vec{e}_z$ is a unit vector for the normal axis.

The blade-hinge frame is the most used for analysis of the blade dynamics.

2.2.6   Position of the blade-hinge frame in the rotor frame can be described by a sequence of two imaginary turns of the rotor frame. The first turn is clockwise (viewing from the below side of the rotor system) around the positive direction of z_r-axis of the rotor frame on the blade azimuthal angle ψ shifted on $\pi/2$ in the counter-clockwise direction: $\pi/2 - \psi$. The second turn is the flapping motion; this turn is clockwise on the blade flapping angle β around the x-axis of the turned frame after the first turn. Finally, the achieved coordinate system is translated from the origin in the hub center (the point O_r) to the intersection of the blade longitudinal axis with the hinge axis (the point O).

If there is an arbitrary vector with defined coordinates in the rotor frame $\vec{W}^{(r)}$, then this vector has coordinates in the blade-hinge frame $\vec{W}^{(h)}$, which are calculated by multiplication of a transformation matrix from the rotor frame to the blade-hinge frame $M^{(r \to h)}$ on this vector in the rotor frame $\vec{W}^{(r)}$:

$$\vec{W}^{(h)} = M^{(r \to h)} \vec{W}^{(r)}.$$

The transformation matrix $M^{(r \to h)}$ can be achieved by multiplication of a transformation matrix of flapping rotation (first multiplicand) on a transformation matrix of azimuthal rotation (second multiplicand):

$$M^{(r \to h)} = \begin{pmatrix} 1 & 0 & 0 \\ 0 & \cos\beta & -\sin\beta \\ 0 & \sin\beta & \cos\beta \end{pmatrix} \begin{pmatrix} \sin\psi & \cos\psi & 0 \\ -\cos\psi & \sin\psi & 0 \\ 0 & 0 & 1 \end{pmatrix},$$

the result is:

$$M^{(r \to h)} = \begin{pmatrix} \sin\psi & \cos\psi & 0 \\ -\cos\beta\cos\psi & \cos\beta\sin\psi & -\sin\beta \\ -\sin\beta\cos\psi & \sin\beta\sin\psi & \cos\beta \end{pmatrix}. \tag{2.1}$$

2.2.7   Orientation of the blade-hinge frame relative to the blade fixed frame can be described by a single clockwise turn of the blade fixed frame around the longitudinal y_b-axis on the blade pitch angle φ. If there is an arbitrary vector with defined coordinates in the blade fixed frame $\vec{W}^{(b)}$, then coordinates of this vector in the blade-hinge frame $\vec{W}^{(h)}$ can be calculated with a transformation matrix from the blade fixed frame to the blade-hinge frame $M^{(b \to h)}$:

$$\vec{W}^{(h)} = M^{(b \to h)} \vec{W}^{(b)},$$

where

$$M^{(b \to h)} = \begin{pmatrix} \cos\varphi & 0 & \sin\varphi \\ 0 & 1 & 0 \\ -\sin\varphi & 0 & \cos\varphi \end{pmatrix}. \tag{2.2}$$

2.2.8   The blade-hinge frame is rotated relative to the inertial coordinate system according to the blade rotations. This blade-hinge frame rotation consists of the rotation around the rotor shaft ($\vec{\omega}$), rotation of the rotor system together with the rotorcraft ($\vec{\Omega}$), and rotation around the blade flapping hinge with angular velocity $-\vec{e}_x d\beta/dt$. Blade-hinge frame angular velocity relative to the inertial coordinate system is represented by a vector $\vec{\omega}_h(\omega_{hx}, \omega_{hx}, \omega_{hx})$ with coordinates in the blade-hinge frame. This angular velocity can be expressed as a vector sum of its components:

$$\vec{\omega}_h = M^{(r \to h)} \left(\vec{\omega} + \vec{\Omega} \right) - \vec{e}_x \frac{d\beta}{dt} = \vec{\omega}^{(h)} + \vec{\Omega}^{(h)} - \vec{e}_x \frac{d\beta}{dt}, \tag{2.3}$$

where $\vec{\omega}^{(h)} = M^{(r \to h)} \vec{\omega}$ is the angular velocity around the rotor shaft with coordinates in the blade-hinge frame, $\vec{\Omega}^{(h)} = M^{(r \to h)} \vec{\Omega}$ is the angular velocity of the rotor system with coordinates the blade-hinge frame. The coordinates of $\vec{\omega}_h$ are determined by coordinates of its components:

$$\begin{aligned} \omega_{hx} &= \Omega_x \sin\psi + \Omega_y \cos\psi - \frac{d\beta}{dt} \\ \omega_{hy} &= \omega \sin\beta - \cos\beta (\Omega_x \cos\psi - \Omega_y \sin\psi) \\ \omega_{hz} &= -\omega \cos\beta - \sin\beta (\Omega_x \cos\psi - \Omega_y \sin\psi) \end{aligned} \tag{2.4}$$

2.3   SWASHPLATE MECHANISM

The swashplate mechanism sets blade pitch angle depending on the position of the swashplate center along the rotor shaft as well as tilt of the swashplate. The swashplate mechanism performs control over the rotor that enables to participate in control over a rotorcraft flight.

2.3.1   The operation of the swashplate mechanism is schematically illustrated according to the scheme in fig. 2.4. The blade pitch angle is set by turning the blade around the feathering hinge together with turning the blade pitch horn AB compelled by motion of the swivel joint (the point B) of the horn with the vertical pitch link BC. The position of the pitch link BC is determined by position of the joint of the link with the rotating swashplate (the point C), which depends on position of the rotating swashplate. The rotating swashplate position is determined by the stationary swashplate position; the rotating swashplate center moves together with the center of the stationary swashplate along the rotor axis (the point O_{SW}); the rotating swashplate is tilted together with the stationary swashplate. Thus, the specified swashplate mechanism impacts the blade pitch angle in two ways: by displacement of the stationary swashplate along the rotor axis that change blade pitch angle at any azimuthal position; by tilt of the stationary swashplate about the swashplate center located on the rotor axis. Besides this, the swashplate mechanism decreases the blade pitch angle at blade flapping upward and increases at flapping downward: so-called flapping compensation.

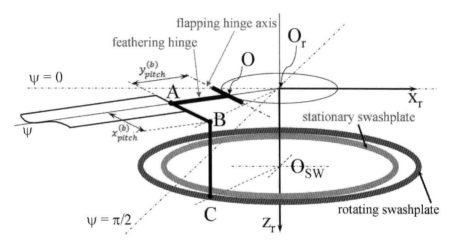

Figure 2.4   Scheme of swashplate mechanism.

It is useful hereafter to represent position of the swivel joint (the point B) of the pitch horn with the pitch link as a radius-vector $\vec{p}_{pitch}^{(b)}(x_{pitch}^{(b)}, y_{pitch}^{(b)}, 0)$ with coordinates in the blade fixed frame, where $x_{pitch}^{(b)} = AB$ is the pitch horn length and $y_{pitch}^{(b)} = AO$ is distance from the pitch horn to the flapping hinge axis (fig. 2.4). The line segment OO' represents the flapping hinge offset: $l_{FH} = OO'$. The distance from the pitch horn to the hub center AO' is the sum of the flapping hinge offset l_{FH} and the distance from the pitch horn to the flapping hinge axis $y_{pitch}^{(b)}$: $AO' = l_{FH} + y_{pitch}^{(b)}$. The distance from the swashplate center (the point O_{SW}) to the joint of the pitch link with the rotating swashplate (the point C) represents the radius of the rotation swashplate and is denoted R_{SW}; usually, this radius is about the distance from the pitch horn to the hub center ($R_{SW} \approx l_{FH} + y_{pitch}^{(b)}$).

2.3.2   Consider a case presented in fig. 2.5, when a swashplate plane is perpendicular to the rotor axis, and a blade lays in the rotor plane ($\beta = 0$). The swashplate plane is parallel to the rotor plane in this case. Initial blade pitch angle is determined by initial position of the swashplate center O_{SW1} on the rotor axis: the center swashplate position determines location of the joint of the pitch link with the rotating swashplates (the point C); the pitch link pulls the joint of the link with the blade pitch horn (the point B); that turns the pitch horn AB together with the blade around its feathering hinge and sets certain initial blade pitch angle. If the stationary swashplate is displaced downward in new swashplate center position O_{SW2}, then: the joint of the pitch link with the rotating swashpalate moves in new position C', where $C'C = O_{SW2}O_{SW1}$; that moves the pitch link BC parallel to the rotor axis downward in new position $B'C'$; that changes the position of the joint of the link with the horn from the point B in a point B', where $B'B = C'C$; that turns the horn together with the blade at angle $\Delta\varphi$ in new position AB'; thus, the blade pitch angle is decreased on $\Delta\varphi$. This pitch angle decrease can be found from a triangle ABB' (fig. 2.5):

$$\sin\Delta\varphi = \frac{BB'}{AB'} = \frac{C'C}{x_{pitch}^{(b)}}.$$

In terms of the swashplate displacement $C'C = O_{SW2}O_{SW1}$, and with assumption of

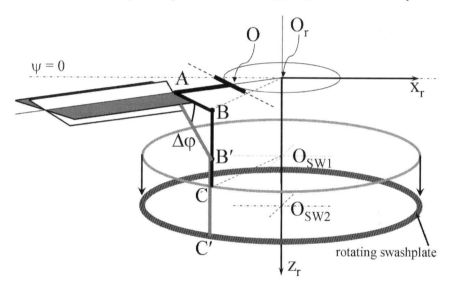

Figure 2.5 Collective pitch.

small pitch angle change that $\sin\Delta\varphi \approx \Delta\varphi$:

$$\Delta\varphi = \frac{O_{SW2}O_{SW1}}{x_{pitch}^{(b)}}.$$

There is some swashplate center position O_{SW0} that causes zero blade pitch angle: $\varphi = 0$ at O_{SW0}. The blade pitch angle φ at certain swashplate center position O_{SW} can be referred to the position O_{SW0}:

$$\varphi = -\frac{O_{SW}O_{SW0}}{x_{pitch}^{(b)}}.$$

Such blade pitch angle determined by the swashplate displacement does not depend on the blade azimuthal position, is called the collective pitch, and is denoted by θ_0:

$$\theta_0 = -\frac{O_{SW}O_{SW0}}{x_{pitch}^{(b)}}.$$

The collective pitch is a basic mechanism to uniformly control all pitch angles of all blades and to control the total lift forces created by the blades.

2.3.3   Here is discussed a case when the swashplate plane is not perpendicular to the rotor axis as illustrated in fig. 2.6. The swashplate plane is tilted in this case and is not parallel to the rotor plane. The radial line segment $O_{SW}C'$, which joins the swashplate center with the joint (the point C') of the rotating swashplate with the pitch link in azimuthal position ψ_{SW}, is tilted to the rotor plane at angle ε_{SW}. The swashplate tilt ε_{SW} in a certain azimuthal position ψ_{SW} is stated to be positive if this joint locates below the swashplate center there. This angle depends on swashplate tilt and on the azimuthal position ψ_{SW} and can be expressed in the following way:

$$\tan \varepsilon_{SW} = \tan \kappa \cos \psi_{SW} + \tan \eta \sin \psi_{SW}, \qquad (2.5)$$

where κ represents swashplate tilt in the zero azimuthal position ($\psi_{SW} = 0$) and is called longitudinal swashplate tilt, η is swashplate tilt in azimuthal position $\pi/2$ and is called lateral swashplate tilt. Here must be noted that azimuthal position of

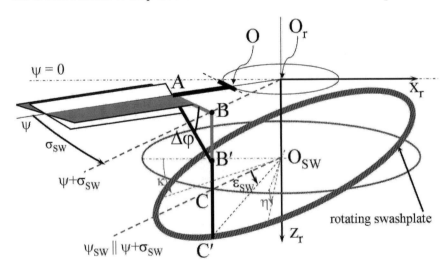

Figure 2.6   Cyclic pitch.

the blade ψ differs from the azimuthal position ψ_{SW} of the joint (the point C') of the rotating swashplate with the pitch link, which is connected to this blade. This joint locates ahead from the blade azimuthal position at angle σ_{SW} toward the shaft rotation. The angle σ_{SW} is determined by the blade horn length AB and the distance from the horn to the rotor axis AO':

$$\sigma_{SW} = \arctan\frac{AB}{AO'} = \arctan\frac{x_{pitch}^{(b)}}{l_{FH}+y_{pitch}^{(b)}}.$$

Let consider that the swashplate initially is not tilted to the rotor axis; the swash-plate center O_{SW} is unchanged on the rotor axis; the blade locates at azimuthal position ψ; the joint of the rotating swashplate with the pitch link (the point C) is correspondingly located in azimuthal position $\psi_{SW} = \psi + \sigma_{SW}$; the pitch link BC sets the pitch horn AB with initial blade pitch angle, which is defined above as the collective pitch θ_0. Swashplate tilt ε_{SW} in ψ_{SW} becomes non-zero after certain turn of the stationary swashplate with longitudinal swashplate tilt κ and lateral swash-plate tilt η. This tilt changes vertical position of the joint of the pitch link with the rotating swashplate from C to C' on $CC' = O_{SW}C'\sin\varepsilon_{SW}$; the pitch link pulls the joint with the pitch horn from the B to B' where $BB' = CC'$; the pitch horn is turned and changes the blade pitch angle on $\Delta\varphi$ (fig. 2.6):

$$\sin\Delta\varphi = -\frac{BB'}{AB'} = -\frac{CC'}{x_{pitch}^{(b)}} = -\frac{O_{SW}C'}{x_{pitch}^{(b)}}\sin\varepsilon_{SW}.$$

If the swashplate is tilted in ψ_{SW} with positive ε_{SW}, then the blade pitch angle is decreased on $|\Delta\varphi|$; therefore, there is minus in front. The line segment $O_{SW}C'$ represents the rotating swashplate radius R_{SW} introduced above. The pitch angle change can be rewritten according to the dependence of $\tan\varepsilon_{SW}$ on κ and η (2.5) assuming, that ε_{SW} is small enough to approximate $\sin\varepsilon_{SW} \approx \tan\varepsilon_{SW}$:

$$\sin\Delta\varphi = -\frac{R_{SW}}{x_{pitch}^{(b)}}\tan\kappa\cos\psi_{SW} - \frac{R_{SW}}{x_{pitch}^{(b)}}\tan\eta\sin\psi_{SW}.$$

It must be taken into account here that pitch angle of the blade in ψ is changed by swashplate tilt in $\psi_{SW} = \psi + \sigma_{SW}$. The assumption of small swashplate tilt ($\tan\kappa \approx \kappa$ and $\tan\eta \approx \eta$) causes small blade pitch angle change $\sin\Delta\varphi \approx \Delta\varphi$. With these:

$$\Delta\varphi = -\frac{R_{SW}}{x_{pitch}^{(b)}}\kappa\cos(\psi+\sigma_{SW}) - \frac{R_{SW}}{x_{pitch}^{(b)}}\eta\sin(\psi+\sigma_{SW}) =$$

$$= -\frac{R_{SW}}{x_{pitch}^{(b)}}(\kappa\cos\sigma_{SW}+\eta\sin\sigma_{SW})\cos\psi - \frac{R_{SW}}{x_{pitch}^{(b)}}(\eta\cos\sigma_{SW}-\kappa\sin\sigma_{SW})\sin\psi$$

or

$$\Delta\varphi = -\theta_1\cos\psi - \theta_2\sin\psi,$$

with
$$\theta_1 := \frac{R_{SW}}{x_{pitch}^{(b)}}(\kappa\cos\sigma_{SW}+\eta\sin\sigma_{SW}) \quad\text{and}\quad \theta_2 := \frac{R_{SW}}{x_{pitch}^{(b)}}(\eta\cos\sigma_{SW}-\kappa\sin\sigma_{SW}),$$

where θ_1 is the lateral cyclic pitch, θ_2 is the longitudinal cyclic pitch, and $\Delta\varphi$ is called the cyclic pitch in the considered case. Thus, the blade pitch angle in azimuthal position ψ is determined by the collective pitch θ_0 and the cyclic pitch:

$$\varphi(\psi) = \theta_0 + \Delta\varphi = \theta_0 - \theta_1 \cos\psi - \theta_2 \sin\psi.$$

The cyclic pitch is a mechanism to control longitudinal and side forces and moments created by the rotor how will be explained further.

2.3.4   The blade pitch angle also changes at blade flapping motion with this swashplate construction as illustrated in fig. 2.7. Consider that the blade locates in azimuthal position ψ, initially lays in the rotor plane and has initial pitch angle φ_θ determined by current swashplate position and tilt: $\varphi_\theta := \theta_0 - \theta_1 \cos\psi - \theta_2 \sin\psi$. If the blade flaps upward with flapping angle $\beta > 0$, then the joint of the pitch horn with the pitch link (the point B) remains unchanged; the point A, where the blade horn axis crosses the blade longitudinal axis, moves upward in new position A', where the displacement $AA' = 2AO\sin(\beta/2)$ is found from an isosceles triangle $AA'O$ with $AO = A'O$. That turns the pitch horn from AB to $A'B$ and changes the blade pitch angle on $\Delta\varphi$, which can be expressed from a isosceles triangle $AA'B$ with $AB = A'B$ (fig. 2.7):

$$2\sin\frac{\Delta\varphi}{2} = -\frac{AA'}{AB} = -2\frac{AO}{AB}\sin\frac{\beta}{2} = -2\frac{y_{pitch}^{(b)}}{x_{pitch}^{(b)}}\sin\frac{\beta}{2}.$$

The minus sign in front indicates blade pitch angle decrease at upward blade flap-

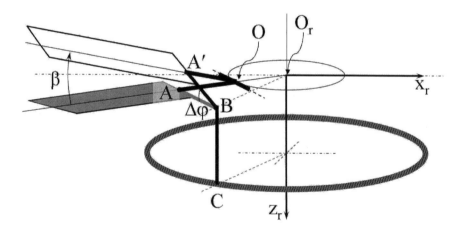

Figure 2.7 Flapping compensation.

ping. Such decrease of the blade pitch angle due to the upward blade flapping is called the flapping compensation and is provided by blade horn attached from the leading edge of the blade: $x_{pitch}^{(b)} > 0$. Assuming the small blade flapping angle

$2\sin(\beta/2) \approx \beta$ that causes the small blade pitch angle change $2\sin(\Delta\varphi/2) \approx \Delta\varphi$:

$$\Delta\varphi = -k\beta,$$

where $k := y^{(b)}_{pitch}/x^{(b)}_{pitch}$ is the flapping compensation coefficient, which describes the flapping compensation and indicates the blade pitch angle changes with the blade flapping. Thus, the blade pitch angle for the discussed case can be expressed in the following way:

$$\phi = \varphi_\theta - k\beta.$$

2.3.5   Summering all defined above, pitch angle φ of a blade in azimuthal position ψ depends on the swashplate position/tilt and on the blade flapping. The blade pitch angle is determined by the collective pitch θ_0, the lateral cyclic pitch θ_1, the longitudinal cyclic pitch θ_2, flapping compensation $k\beta$:

$$\varphi(\psi) = \theta_0 - \theta_1 \cos\psi - \theta_2 \sin\psi - k\beta. \tag{2.6}$$

Hereafter, the swashplate impact on the blade pitch angle will be described only by the collective pitch θ_0, the lateral cyclic pitch θ_1, and the longitudinal cyclic pitch θ_2. The longitudinal κ and lateral η swashplate tilts are important for the rotorcraft flight dynamics; however, these parameters are avoided in the analysis of the rotor properties presented here.

2.4   GENERAL STATEMENTS OF THE BLADE ELEMENT ROTOR THEORY

2.4.1   The blade element rotor theory considers each rotor blade as a set of imaginary elements, into which the blade can be transversely divided. Each blade element is considered as a wing with certain aerodynamic properties. Length of an element is assumed to be small enough so that the element is homogeneously blown by any airflow. Position of a certain blade element in the rotor frame can be described by azimuthal position of its blade ψ, by the blade flapping angle β, and by distance y_b from the flapping hinge axis to the element middle point along the blade longitudinal y_b-axis.

2.4.2   It is agreed hereafter, in order to avoid sign misunderstanding, that all element velocity terms are specified as a motion of a blade element relative to its oncoming airflow. It is stated here as well that all used angular terms have radian units, all used angular velocity terms have radian per second units.

2.4.3   The rotor shaft is rotated under engine torque; the rotor shaft rotates the rotor hub together with flapping hinges, to which blades are attached; thus, the blades are rotated together with the rotor shaft. The rotated blades are affected by centrifugal forces, which compel the blades to rotate close to the rotor plane.

2.4.4   During the rotor shaft rotation together with the blades in the undisturbed airspace, each blade element is blown by oncoming airflow in direction of the blade rotational motion. Elements with different distances to the rotor axis are blown by different velocities of oncoming airflows due to circumferential motion around the rotor shaft: an element, which is far from the rotor axis, is blown by the airflow with greater velocity.

In a case of forward motion of the whole rotor system in the undisturbed airspace ($V_x \neq 0$), a blade is additionally blown by oncoming airflow due to the forward motion, which is different at different blade azimuthal positions: the oncoming airflow is greater at blade azimuthal position $\psi = \pi/2$ (the advancing blade); the oncoming airflow is weaker at blade azimuthal position $\psi = 3\pi/2$ (the retreating blade). The airflow due to the forward motion is directed along the blade longitudinal y-axis at position $\psi = 0$ (the back blade) and in the opposite direction to this axis at $\psi = \pi$ (the front blade); only transverse airflow is caused by blade rotational motion at $\psi = 0$ and $\psi = \pi$.

In a case of vertical motion of the rotor system ($V_y \neq 0$), each blade element is blown by the same vertical airflow velocity along the z_r-axis of the rotor frame.

2.4.5   During rotation of the blades in the airspace, vortex sheets are created in such a way, that the air is sucked from above of the rotor system and is blown out downward. This additional sucked air flows through the rotor disc and is called the induced airflow hereafter. The induced airflow accompanies lift force generation by the blades: the stronger are the generated lift forces, the greater is velocity of the induced airflow. The induced airflow participates in the blowing of each blade element. Generally, the induced airflow velocity is not the same for all blade elements. In order to describe the induced airflow, which blows a blade element at blade azimuthal position ψ and element position y_b along the blade longitudinal axis, there is introduced the induced velocity $V_i(y_b, \psi)$, which describes vertical velocity of the blade element relative to the induced airflow (by the convention) and is directed parallel to the z_r-axis of the rotor frame. For instance, if the induced airflow blows downward, then the blade moves upward relative to this induced airflow; therefore, the induced velocity is negative.

In the frame of the blade element rotor theory, the induced velocity $V_i(y_b, \psi)$ is assumed to be known for any blade element position y_b and ψ.

2.4.6   Air blowing of a blade element generates aerodynamic forces acting on this element. Aerodynamic forces of a blade element consist of the element lift force and the element profile (parasitic) drag. The element lift force is directed perpendicular to the element oncoming airflow velocity and to the blade longitudinal axis. The profile drag is directed along the element oncoming airflow velocity: this is opposite to the element velocity relative to the oncoming airflow. The element aerodynamic force components proportionally depend on dynamic pressure of the element oncoming airflow and the element planform area with proportionality coefficients: the element lift coefficient C_{Le} for the element lift force, and the element profile drag coefficient C_{De} for the element profile drag. It is assumed that: an element lift coefficient depends only on attack angle of the element in its oncoming airflow and this dependence is increasing linearly with a lift coefficient slope C_{Le}^α; and an element profile drag coefficient is constant. It is assumed here that an element does not create any lift force if the element has zero attack angle.

2.4.7   Here, the general case with non-uniform blade aerodynamic properties along the blade longitudinal axis is discussed: this means that the aerodynamic properties may not be identical for all blade elements. If y_b denotes certain position of a blade element on the blade longitudinal axis, then the element lift coefficient slope $C_L^\alpha(y_b)$, the element profile drag coefficients $C_D(y_b)$, and length of the element airfoil chord $c(y_b)$ may vary along the blade longitudinal axis.

2.4.8   Pitch (incidence) angle of a blade element at y_b position along the blade longitudinal axis is defined as the angle on which the element should be turned around the longitudinal axis in such a way, that the element chord would be parallel to the rotor plane. The element pitch angle is denoted by $\varphi_e(y_b)$. Element pitch angle equals zero if the element chord is parallel to the rotor plane; the element pitch angle is positive if the chord front part is directed upward of the rotor system, and is negative if the chord front part is directed downward. The element aerodynamic properties strongly depend on the element pitch angle.

The general case supposes that a blade has twisting: pitch angle of an element at a certain position y_b along the blade may differ on angle $\varphi_{tw}(y_b)$ in comparison to pitch angle of the blade root element, so that $\varphi_e(y_b) = \varphi_e(0) + \varphi_{tw}(y_b)$. The pitch angle of the whole blade is determined by the pitch angle of the blade root element as defined above: $\varphi = \varphi_e(0)$. Since the blade is rigid, the pitch angle of the blade element at blade azimuthal position ψ and at position y_b of the element along the blade is determined by the blade pitch angle, which the swashplate mechanism sets according to (2.6) and by the blade twisting:

$$\varphi_e(y_b, \psi) = \theta_0 + \varphi_{tw}(y_b) - \theta_1 \cos \psi - \theta_2 \sin \psi - k\beta. \tag{2.7}$$

2.4.9   Blade pitch angle is set by turning the blade around its feathering hinge compelled by the blade pitch horn, which joins the vertical pitch link via the swivel joint (fig. 2.8). Position of this swivel joint is described by a radius vector $\vec{p}_{pitch}(x_{pitch}^{(b)}, y_{pitch}^{(b)}, 0)$ with coordinates in the blade fixed frame as introduced in 2.3.1. The pitch link affects the blade by a force, which acts on the blade pitch horn in the swivel joint, and is called the blade pitch force R_{pitch}. Originally, the pitch force is reaction of the pitch link on the blade pitch horn in the swivel joint. The blade pitch force is transferred on the whole rigid blade. This force with the arm $x_{pitch}^{(b)}$ creates a moment around the blade longitudinal y_b-axis, which participates in blade rotation around the blade feathering hinge. Thus, the swashplate mechanism applies via the pitch link a blade pitch force, which is necessary to set blade pitch angle φ in correspondence to the swashplate operation as described in the section 2.4.

According to the Newton's third law, the blade pitch horn acts on the pitch link with a force $\vec{R}_{SP}$, which equals by magnitude to the pitch force $\vec{R}_{pitch}$, with which the pitch link acts on the pitch horn, but acts in the opposite direction: $\vec{R}_{SP} = -\vec{R}_{pitch}$ (fig. 2.8). Originally, this force $\vec{R}_{SP}$ is the reaction force of the blade pitch horn applied to the pitch link in the swivel joint. The force $\vec{R}_{SP}$ is transferred to the swashplate via the pitch link and on the whole rotorcraft. The force $\vec{R}_{SP}$ is called the blade force on the swashplate hereafter. It is assumed hereafter that this force on the swashplate is negligibly small in comparison with other aerodynamic forces created by the blade.

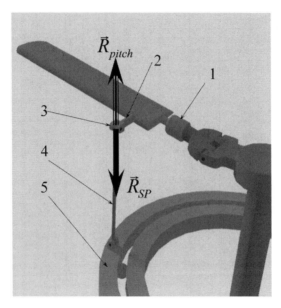

Figure 2.8 Blade pitch force vs blade force on swashplate: 1 – feathering hinge; 2 – blade pitch horn; 3 – swivel joint; 4 – pitch link; 5 – rotating swashplate.

2.4.10   Each blade element is affected by element aerodynamic forces caused by element air blowing, as was described above. Besides this, each blade element is affected by the element weight, centrifugal forces, and Coriolis forces, which are caused by element rotational motions around the rotor axis with flapping motion and rotation of the whole rotor system.

The forces of all blade elements act on their whole blade. Since the blade is assumed an absolutely rigid body, it is not bent nor twisted under these forces. The blade performs flapping motion around its flapping hinge under superposition of the forces of all its elements.

2.4.11   It is simplistically considered here that a blade root part has two separate lugs (numbered 1 and 2), which are put on the blade flapping hinge axle fixed to the rotor hub (fig. 2.9(a)). The blade lugs enable the blade to turn without friction around the axle, the axis of which is the hinge axis. The lugs are located on the axle at certain distance from each other. Mechanical stoppers on the axle are used to avoid the lugs to slip along the axle. The blade interacts with the rotor hub through the interaction of the lugs with the flapping hinge axle and the stoppers.

The rotational motion of the blade around the rotor shaft is ensured by the inter-action of the blade lugs with the flapping hinge axle and the slip stoppers. The axle acts on a i-th lug ($i = 1, 2$) with a reaction force, which is directed perpendicular to the axle and is described by a vector with just longitudinal $R^{(i)}_{Yaxle}$ and normal $R^{(i)}_{Zaxle}$ coordinates in the blade-hinge frame. A stopper acts on the i-th lug with a reaction

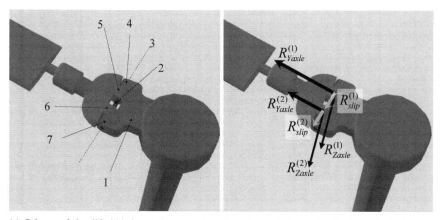

(a) Scheme of simplified blade attachment to flap- (b) Scheme of forces affecting blade by its flapping
ping hinge   hinge

Figure 2.9 Attachment of blade to its flapping hinge: 1 – flapping hinge; 2 – flapping hinge axle; 3 – flapping hinge axis; 4 - slip stopper; 5 – blade lug №1; 6 – blade lug №2; 7- slip stopper.

force $R_{slip}^{(i)}$ directed along the axle coincided with the tangential axis of the blade-hinge frame (fig. 2.9(b)). Thus, each lug is affected by a flapping hinge reaction created by the axle and the stoppers with a force $\vec{R}_{FH}^{(i)}(R_{slip}^{(i)}, R_{Yaxle}^{(i)}, R_{Zaxle}^{(i)})$, which has coordinates in the blade-hinge frame. A flapping hinge reaction force on each lug is transferred on the whole blade as the rigid body.

Cumulative interaction of lugs of a blade with its flapping hinge is conveniently considered about the flapping hinge center as a reference point, which is the blade-hinge frame origin (the point O). Based on this, the flapping hinge affects its blade relative to this reference point by:

- the hinge axle reaction force $R_{Yaxle} = R_{Yaxle}^{(1)} + R_{Yaxle}^{(2)}$ along the longitudinal axis;
- the hinge axle reaction force $R_{Zaxle} = R_{Zaxle}^{(1)} + R_{Zaxle}^{(2)}$ along the normal axis;
- the hinge slip stopper reaction force $R_{slip} = R_{slip}^{(1)} + R_{slip}^{(2)}$ along the tangent axis;
- the force moment $M_{FH} = x_{lug}^{(1)} R_{Yaxle}^{(1)} + x_{lug}^{(2)} R_{Yaxle}^{(2)}$ due to the separate location of the lugs on the hinge axle, where $x_{lug}^{(1)}$ and $x_{lug}^{(2)}$ are positions of the first and second lugs along the tangential axis.

The moment M_{FH} is called the flapping hinge reaction moment and acts on the blade around the normal z-axis of the blade-hinge axis. The reaction forces of the axle and the slip stoppers create no force moments around the x-tangential and y-longitudinal axes due to no friction in the flapping and feathering hinges. The axle reaction forces (R_{Yaxle}, R_{Zaxle}) and the slip stopper reaction force (R_{slip}) compose coordinates of the flapping hinge reaction force $\vec{R}_{FH}(R_{slip}, R_{Yaxle}, R_{Zaxle})$ in the blade-hinge frame, which acts on the blade and is applied to the origin of the blade-hinge frame. The hinge reaction force and the hinge reaction moment fully represent the action of the flapping hinge on its blade.

From another hand, a blade affects the rotor hub through its flapping hinge. According to the Newton's third law, the blade acts on the flapping hinge with a force, which equals by magnitude but acts in opposite direction to the force $\vec{R}_{FH}$, with which the flapping hinge acts on the blade. So that, the blade acts on the rotor hub with a force $\vec{F}_{FH}$ applied to the flapping hinge, which equals $\vec{F}_{FH} = -\vec{R}_{FH}$. This force is called the blade force on the flapping hinge hereafter. The forces, with which the blade acts on the flapping hinge axle through the two separated blade lugs, create a force moment M_{onFH}, which is opposite to the hinge reaction moment $M_{onFH} = -M_{FH}$, is directed around the normal z-axis and is called the blade moment on the flapping hinge. The blade force and moment on the flapping hinge represent the action of the blade on its flapping hinge.

This simplified approach can be applied to any complicated construction of a flapping hinge with a free flapping blade.

2.4.12   The total aerodynamic and gravity forces of all elements of a blade are transferred on its flapping hinge through the interaction of the blade lugs with the flapping hinge axle and the stoppers. The blade force on the flapping hinge is transferred to the rotor hub and the whole rotorcraft.

2.4.13   At certain conditions, a force on a flapping hinge of a single blade might be different at different azimuthal positions and might change during one full turn of the rotor shaft. Such a blade creates vibrations on the rotorcraft with frequency equal to the frequency of the shaft rotation $\omega/(2\pi)$. A next blade repeats the motion of its previous blade and creates a similar force on its flapping hinge with delay, which is n (the number of rotor blades) times shorter than the one rotor full turn. So, all n blades create vibration with frequency, which equals $n\omega/(2\pi)$.

2.4.14   The blade element rotor theory neglects this kind of vibration. However, the forces on all flapping hinges are averaged over one full turn of the rotor shaft. This averaged force is considered about the rotor hub center as a reference point and is called the total rotor force. So, the total rotor force is applied to the hub center and is described by a vector $\vec{R}$ with coordinates in the rotor frame. The total rotor force is transferred to the whole rotorcraft.

The total rotor force is decomposed into three components.

– The rotor thrust is a projection of the total rotor force on the reverse direction of the z_r-axis of the rotor frame and is denoted by T. The rotor thrust is positive if this component is directed upward of the rotor system.
– The rotor longitudinal force is a projection of the total rotor force on the reverse direction of the x_r-axis of the rotor frame and is denoted by H. This component is positive if it is directed opposite to the rotor system forward motion. The positive rotor longitudinal force coincides with zero azimuthal position ($\psi = 0$).

— The rotor side force is a projection of the total rotor force on the y_r-axis of the rotor frame and is denoted by S. The positive rotor side force is directed to the right along azimuthal position $\psi = \pi/2$ $(90°)$.

2.4.15   A blade force on a flapping hinge creates a moment about the hub center due to the flapping hinge offset. Forces on the flapping hinges of all rotor blades create the total moment about the hub center, which is averaged over one shaft turn in order of generalization. In a case of different flapping hinge forces at different azimuthal positions, the average total moment over one rotor shaft full turn is not zero. The averaged total moment is decomposed into two components: lateral hub moment M_{HUBx} around the rotor longitudinal x_r-axis of the rotor frame; and longitudinal hub moment M_{HUBy} around the rotor lateral y_r-axis of the rotor frame. This moment is transferred on the rotorcraft concerning the rotor hub center as the reference point.

2.4.16   During rotation of the rotor blades around the rotor shaft, blade elements create tangential forces, which consist of aerodynamic drag. These forces act in the reverse direction of the rotational motion of the elements around the rotor shaft, create the resistance moment opposite to this rotation and promote the rotor to slow down. The rotor resistance moment represents a sum of the resistance moments of all elements of all blades and is directed around the z_r-axis of the rotor frame. To permanently keep the blade rotation constant, engine torque is needed to be applied to the rotor shaft, which equals the rotor resistance moment by magnitude but acts in opposite direction; the engine torque counteracts the rotor resistance torque. In this case, the engine torque action creates a reactive moment on the rotorcraft, which is opposite to the engine torque and is the same as the resistance torque. This reactive moment is called the rotor torque, is denoted by Q, and is applied to the rotorcraft around the z_r-axis of the rotor frame.

2.4.17   The steady rotation of rotor blades around the rotor shaft means that every blade repeats identical flapping motion at same azimuthal position every turn around the rotor shaft. This condition ensures that the total rotor force remains constant and is not affected by a transient state. Only the steady rotation of a blade is discussed hereafter.

2.4.18   Analysis and understanding of the physical phenomenon of thrust creation by such rotor systems are systematized in the blade element rotor theory. This theory is based on principles of classical (Newtonian) mechanics, rigid body mechanics, and aerodynamics. The main goal of the blade element rotor theory is to determine all components of the total rotor force, the total rotor moment, and the rotor torque according to rotorcraft flight conditions and control position, which are determined by the swashplate tilt and position. This can be done with the understanding of the flapping motion, which is analyzed as well in the frame of the blade element rotor theory.

2.5   BASIC ASSUMPTIONS AND APPROXIMATIONS

In the frame of the blade element rotor theory, it is common to do some assumptions and approximations, which neglect effects, which are orders of magnitude smaller than basic effects, to simplify the analysis of basic properties of the rotor system and to get possible analytical solutions. Here are specified general assumptions, which are common for the discussed and the conventional blade element rotor theory.

2.5.1 It is assumed that the rotor system carries its rotorcraft with a weight, which is much greater than the total weight of all rotor blades. So, the rotor blades create during a rotorcraft flight the forces acted on the rotorcraft, which are much greater than their own weight. Therefore, the aerodynamic forces of any blade element surpass the weight of this element, and the element weight can be neglected.

2.5.2 It is assumed that the rotorcraft weight is small enough to be carried by the rotor system in the airspace but is great enough to move with acceleration, which is insignificant enough to neglect the inertial forces caused by this acceleration.

2.5.3 The rotorcraft performs maneuvres with rotorcraft rotations, which are much slower than the rotor shaft rotation: $|\Omega_x| \ll \omega$ and $|\Omega_y| \ll \omega$.

2.5.4 It is assumed that the blade flapping angle (β) at any condition is small enough to approximate $\sin\beta \approx \beta$ and $\cos\beta \approx 1$. It is the most applied approximation, especially for a mathematical derivation of blade flapping motion.

2.5.5 It is assumed that blade element circumferential velocity due to the rotation around the rotor shaft $r\omega$ is dominant, where $r = l_{FH} + y_b \cos\beta$ is the distance from the middle point of the element to the rotor axis. This assumption is not strict for blade elements close to the rotor shaft; however, blade elements in this area do not strongly affect the dynamic properties of the rotor system due to small blade element velocities.

If velocity of a blade element is represented by a vector in the blade-hinge frame with coordinates $\vec{V}_e(V_{ex}, V_{ey}, V_{ez})$, then the tangential coordinate V_{ex}, which contains the circumferential component of the rotation around the rotor shaft ($r\omega$), dominates in comparison to other coordinates: $V_{ex} \gg |V_{ey}|$ and $V_{ex} \gg |V_{ez}|$. A consequence of the assumption is that the magnitude of the element velocity can be approximated by the V_{ex} coordinate: $|\vec{V}_e| \approx V_{ex}$.

2.5.6 It is assumed that any blade pitch angle is small enough to approximate $\sin\varphi \approx \varphi$ and $\cos\varphi \approx 1$. It is assumed as well that angle of attack of any blade element is less than stalling angle.

2.5.7 It is assumed that a moment of blade inertia around the blade longitudinal y_b-axis ($I_{yy}^{(b)}$) is much smaller than moments of blade inertia around the blade transverse x_b-axis ($I_{xx}^{(b)}$) and the blade normal z_b-axis ($I_{zz}^{(b)}$): $I_{yy}^{(b)} \ll I_{xx}^{(b)}$ and $I_{yy}^{(b)} \ll I_{zz}^{(b)}$. This assumption is based on the high wing aspect ratio of the blade. The consequence of this assumption is that these moments of blade inertia around the blade transverse and blade normal axes are approximately equal: $I_{xx}^{(b)} \approx I_{zz}^{(b)}$.

2.5.8 It is assumed that the projection of the hub center velocity on the rotor plane changes very slow its orientation relative to the inertial coordinate system in comparison to the rotor shaft angular velocity. It means that the rotation of the rotor frame around its z_r-axis in the inertial coordinate system is negligible. This assumption ensures the relationship $d\psi/dt = \omega$.

2.6 COMPARISON WITH THE CONVENTIONAL THEORY

The blade element rotor theory, which is presented hereafter, is based on the "classical" conventional blade element rotor theory for helicopter rotors, which was established during the 20-th century. Referencing to the conventional blade element rotor theory within the issue means the references to the sources [4][3][7][2][5][6][1]. The great part of the further discussion states the basics of this conventional theory. The challenge of the discussed blade element rotor theory is to enhance calculations of the total rotor force, the rotor moments, and the rotor torque according to the requirements of helicopter flight simulations. The calculations, which can be performed without iterative algorithms but with direct computation, are preferred. The basic differences of the current theory with the conventional theory are presented here.

2.6.1 The conventional theory assumes that the aerodynamic properties of a blade element such as lift coefficient slope, drag coefficient, and chord, are same for any element of the blade: $C_L^\alpha(y_b) = const$; $C_D(y_b) = const$; and $c(y_b) = const$. The current theory assumes that these parameters can differ for any blade element. This change of the aerodynamic properties along the blade is generalized in the frame of the current theory.

2.6.2 The conventional theory assumes that a blade has no twisting: $\varphi_{tw}(y_b) = 0$. The current theory analyzes and generalizes any kind of gradual change of blade element pitch angle along a blade.

2.6.3 The conventional theory assumes that induced airflow is homogeneous around the rotor disc: $V_i(y_b, \psi) = const$. The current theory analyzes impact of the inhomogeneity of the induced airflow velocity on the rotor system properties with a caveat that the induced velocity is known at any point of the rotor disc.

2.6.4 The conventional theory assumes that blade flapping hinge offset l_{FH} is short enough in comparison with the rotor radius R to be neglected. Despite the small value of the blade flapping hinge offset, the current theory analyzes the influence of this offset on output characteristics.

2.6.5 The conventional theory neglects the angular acceleration of the rotor system rotation $(d\vec{\Omega}/dt)$. The current theory analyzes the impact of such angular acceleration despite its small value.

2.6.6 Concepts of a tip path plane and a plane of constant pitch angle, as well as a concept of an equivalent rotor, are avoided in the current theory.

3 Blade Element Properties

The essential idea of the blade element rotor theory is that a rotor blade is considered as a set of transversely divided elements, each of which is assumed as a wing. Each blade element as a wing creates an aerodynamic lift force in oncoming airflow. A blade element has small enough length to ensure homogeneous aerodynamics properties and homogeneous air blowing conditions. The homogeneous element aerodynamic properties mean that the element chord length c_e and element aerodynamic coefficients (C_{Le}^{α} and C_{De}) are constant along the element. The homogenous blowing conditions mean that the air velocity at each point of the element is the same and equals the element velocity relative to the airflow.

Dynamic and aerodynamic properties of a blade element are discussed in the chapter in terms of their further generalization within a whole rigid blade.

3.1 BLADE ELEMENT DESCRIPTION

3.1.1   A middle point of a blade element is defined here as a point on the blade longitudinal y_b-axis with equal distance to the transverse edges of this element. A middle point of a blade element is supposed to be a mass center of the element as a point, which lays on the blade principal axis of inertia.

3.1.2   A blade element position on its blade is defined by position of the element middle point on the blade longitudinal y_b-axis of the fixed blade frame. The element length is defined by distance between the element transverse edges along the blade longitudinal y_b-axis and is denoted by dy_b.

It is as well useful to define a blade element position in the rotor frame (fig. 3.1): by azimuthal position of its blade (ψ); by the blade flapping angle (β); and by position of the element along the blade (y_b). The section radius of the element r is defined as a distance from the blade element middle point to the rotor axis: $r = l_{FH} + y_b \cos \beta$.

3.1.3   Position of a blade element in the rotor frame is represented by a blade element radius-vector $\vec{r}$, which joins the rotor frame origin (the hub center O_r) and the blade element middle point. This vector has coordinates in the rotor frame $\vec{r}(-r\cos\psi,\ r\sin\psi,\ -(r - l_{FH})\tan\beta)$.

3.1.4   A chord of a blade element (c_e) is defined as a transverse line segment passing the element middle point and joins leading and trailing edges of the blade element. It is assumed that if the oncoming airflow is directed along the blade element chord, then the blade element attack angle equals zero, and the blade element does not create any lift force in this airflow.

3.1.5   Planform area of a blade element is defined as an area of the blade element projection on the $x_b y_b$-plane of the blade fixed frame and is denoted by dS. Due to the homogenous properties of the blade element, the element planform equals the element chord multiplied on the element length: $dS = c_e dy_b$.

3.1.6   A blade element has a mass, which is denoted by dm.

DOI: 10.1201/9781003296232-3

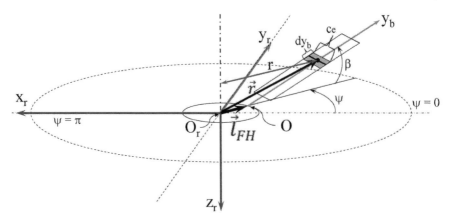

Figure 3.1   Scheme of the rotor system in terms of blade element.

3.1.7   A blade element is considered here as an absolutely rigid body, which is a part of an absolutely rigid blade.

3.1.8   Central principal inertia axes of a blade element are fixed to the blade fixed frame and pass the element middle point. One central principal axis is aligned along the blade longitudinal axis; another is aligned to the element chord; a third is perpendicular to the previous two.

3.2   ABSOLUTE VELOCITY OF A BLADE POINT

Each blade moves relative to the inertial coordinate system, which is stated here to be the undisturbed airspace. Different points of the blade are moving differently because of blade rotational motion. The velocity of a blade point is considered here relative to the inertial coordinate system. For further purposes, such absolute velocities are represented here with coordinates in the blade-hinge frame.

3.2.1   There is a blade point, which is fixed to a blade. The position of the point relative to the blade is described by a radius-vector with coordinates in the blade fixed frame $\vec{p}^{(b)}(x_b, y_b, z_b)$. This radius-vector joins the flapping hinge center (the point O in fig. 3.1) with the blade point. These coordinates are unchanged because the point is fixed to the blade.

Position of the blade point in the blade-hinge frame is described by a radius-vector $\vec{p}$. Since the longitudinal y-axis of the blade-hinge frame coincides with the y_b-axis of the blade fixed frame, then the y-coordinate of $\vec{p}$ equals the y_b-coordinate of the $\vec{p}^{(b)}$, and remains unchanged. The $x-$ and $z-$ coordinates can change due to the blade pitch angle change. Knowing the transformation matrix from the blade fixed frame to the blade-hinge frame $M^{(b \to h)}$ (2.2), the coordinates of the radius-vector $\vec{p}(x, y_b, z)$ can be found as:

$$\vec{p} = M^{(b \to h)} \vec{p}^{(b)}.$$

Position of the blade point in the rotor frame is described by a radius-vector $\vec{r}$, which joins the rotor hub center (the point O_r in fig. 3.1) with the blade point. Because of the flapping hinge offset, this radius-vector equals to a vector sum of the flapping hinge offset $\vec{l}_{FH}^{(r)}$ in the rotor frame and the radius-vector $\vec{p}^{(r)}$ from the flapping hinge center to the blade point in the rotor frame:

$$\vec{r} = \vec{l}_{FH}^{(r)} + \vec{p}^{(r)},$$

where $\vec{l}_{FH}^{(r)}$ is the radius-vector from the hub center (the point O_r) to the center of the flapping hinge (the point O) with coordinates in the rotor frame. Knowing the transformation matrix $M^{(r \to h)}$ (2.1) from the rotor frame to the blade-hinge frame, the relationship between $\vec{p}$ and $\vec{r}$ is:

$$M^{(b \to h)}\vec{r} = \vec{l}_{FH}^{(h)} + \vec{p},$$

where $\vec{l}_{FH}^{(h)}(0, l_{FH}\cos\beta, l_{FH}\sin\beta)$ is a vector of the flapping hinge offset in the blade-hinge frame.

3.2.2 The blade point absolute velocity measures the velocity of the blade point relative to the inertial coordinate system and is represented by a vector with coordinates in the blade-hinge frame $\vec{v}(v_x, v_y, v_z)$. The blade point absolute velocity is the vector sum of the following components.

- The velocity of the blade point caused by motion of the whole rotor system relative to the inertial coordinate system. This velocity is described by the hub center velocity with coordinates in the rotor frame $\vec{V}(V_x, 0, V_z)$ as defined in 2.2.1. Knowing the transformation matrix from the rotor frame to the blade-hinge frame $M^{(r \to h)}$, this velocity in the blade-hinge frame has coordinates:

$$M^{(r \to h)}\vec{V} = (V_x\sin\psi, \ -V_x\cos\beta\cos\psi - V_z\sin\beta, \ -V_x\sin\beta\cos\psi + V_z\cos\beta).$$

- The circumferential velocity of the blade point due to the rotation of its blade around the rotor shaft. This velocity is always parallel to the rotor plane and perpendicular to the radius-vector $\vec{r}$ from the rotor frame origin to the blade point. This velocity component can be represented in a vector form as a cross-product $\vec{\omega} \times \vec{r}$ in the rotor frame, and as $M^{(r \to h)}(\vec{\omega} \times \vec{r}) = \vec{\omega}^{(h)} \times (\vec{l}_{FH}^{(h)} + \vec{p})$ in the blade-hinge frame, where $\vec{\omega}^{(h)} = M^{(r \to h)}\vec{\omega}$ is the angular velocity around the rotor shaft with coordinates in the blade-hinge frame. This velocity component of a blade point, which lays on the blade longitudinal axis, has the magnitude $r\omega$, is always directed along the tangential x-axis of the blade-hinge frame, and has coordinates $(r\omega, 0, 0)$ in the blade-hinge frame, where r is the distance from the blade point to the rotor axis.
- The blade point velocity due to the blade flapping motion is determined by the circumferential velocity of the blade point during turning of the blade around its flapping hinge. The angular velocity of the flapping motion is defined by

the rate of flapping angle change $d\beta/dt$. A vector of the angular velocity is directed along the tangential x-axis of the blade-hinge frame, which coincides with the flapping hinge axis, and can be expressed as $-\vec{e}_x d\beta/dt$; the minus sign is explained by clockwise rotation at increasing of the flapping angle. The circumferential velocity of the blade point can be found by a cross-product $-(\vec{e}_x \times \vec{p})d\beta/dt$.

— The blade point circumferential velocity due to rotation of the rotor system together with the rotorcraft. This rotation is represented by the angular velocity $\vec{\Omega}(\Omega_x, \Omega_y, 0)$ in the rotor frame as defined in 2.2.1. The vector of the circumferential velocity can be expressed as a cross-product $\vec{\Omega} \times \vec{r}$ in the rotor frame, and as $M^{(r \to h)}(\vec{\Omega} \times \vec{r}) = \vec{\Omega}^{(h)} \times (\vec{l}_{FH}^{(h)} + \vec{p})$ in the blade-hinge frame, where $\vec{\Omega}^{(h)} = M^{(r \to h)}\vec{\Omega}$ is the angular velocity of the rotor system with coordinates in the blade-hinge frame.

— The blade point circumferential velocity due to rotation around the feathering hinge. The angular velocity of this rotation is defined by a rate of blade pitch angle change $d\varphi/dt$. The increase in the blade pitch angle causes counter-clockwise rotation around the longitudinal y-axis of the blade-hinge frame. The angular velocity can be represented by a vector along the y-axis: $\vec{e}_y d\varphi/dt$. The circumferential velocity of the blade point due to the pitch angle change can be found as a cross-product $(\vec{e}_y \times \vec{p})d\varphi/dt$. If the blade point lays on the longitudinal axis ($\vec{p}(0, y_b, 0)$), then this circumferential velocity equals zero.

Having all components, the blade point absolute velocity with position $\vec{p}$ can be expressed in a vector form:

$$\vec{v} = M^{(r \to h)}\vec{V} + \left(\vec{\omega}^{(h)} + \vec{\Omega}^{(h)}\right) \times \left(\vec{l}_{FH}^{(h)} + \vec{p}\right) + \left(-\frac{d\beta}{dt}\vec{e}_x + \frac{d\varphi}{dt}\vec{e}_y\right) \times \vec{p}. \quad (3.1)$$

Such blade point absolute velocity is used in dynamic properties analysis of a blade element.

3.2.3 Absolute velocity of a blade element is defined as the absolute velocity of the element middle point, which is located on the blade longitudinal axis and is stated to be the center of mass of the element. The element middle point is described by a radius-vector with coordinates in the blade-hinge frame $\vec{p}_e(0, y_b, 0)$. The element absolute velocity is not affected by rotation around the feathering hinge because of the location of the element middle point on the longitudinal axis. The element absolute velocity is described by a vector with coordinates in the blade-hinge frame $\vec{v}_e(v_{ex}, v_{ey}, v_{ez})$ and can be expressed based on (3.1):

$$\vec{v}_e = M^{(r \to h)}\vec{V} + \left(\vec{\omega}^{(h)} + \vec{\Omega}^{(h)}\right) \times \left(\vec{l}_{FH}^{(h)} + y_b\vec{e}_y\right) - y_b\frac{d\beta}{dt}\vec{e}_x \times \vec{e}_y$$

with coordinates:

$$v_{ex} = (y_b \cos\beta + l_{FH})\omega + V_x \sin\psi + y_b \sin\beta(\Omega_x \cos\psi - \Omega_y \sin\psi),$$
$$v_{ey} = -V_x \cos\beta \cos\psi - \sin\beta(V_z + l_{FH}(\Omega_x \sin\psi + \Omega_y \cos\psi)),$$
$$v_{ez} = -V_x \sin\beta \cos\psi + V_z \cos\beta + (y_b + l_{FH}\cos\beta)(\Omega_x \sin\psi + \Omega_y \cos\psi) - y_b \frac{d\beta}{dt}.$$
$$(3.2)$$

3.3   BLADE ELEMENT AIR VELOCITY

3.3.1   A blade element is blown by oncoming airflow. Since the airflow is as-sumed homogenous for the whole blade element, then the oncoming airflow velocity is the same for all parts of the blade element. The blade element air velocity is defined as the velocity of the blade element middle point relative to the oncoming airflow. The blade element air velocity is represented by coordinates in the blade-hinge frame $\vec{V}_e(V_{ex}, V_{ey}, V_{ez})$. The element airspeed is defined as the magnitude of the element air velocity and is denoted as $V_e := |\vec{V}_e|$.

The element air velocity is important to determine aerodynamic properties of the blade element.

3.3.2   Since the undisturbed airspace is stated here as the inertial coordinate system, then the element air velocity equals to the sum of the element absolute ve-locity and the induced velocity, which is caused by perturbations created by the rotor system itself. It is assumed that the induced velocity is directed strictly parallel to the rotor axis. The induced velocity, which blows an element located at y_b on its blade at blade azimuthal position ψ, can be described by a vector with coordinates in the rotor frame $\vec{V}_i(0, 0, V_i(y_b, \psi))$. This induced velocity has coordinates in the blade-hinge frame:

$$M^{(r \to h)}\vec{V}_i = \left(0, \quad -V_i(y_b, \psi)\sin\beta, \quad V_i(y_b, \psi)\cos\beta\right).$$

3.3.3   With these induced velocity coordinates and the element absolute velocity coordinates, which were found above (3.2), the element air velocity $\vec{V}_e$ has coordi-nates in the blade-hinge frame:

$$V_{ex} = (y_b \cos\beta + l_{FH})\omega + V_x \sin\psi + y_b \sin\beta(\Omega_x \cos\psi - \Omega_y \sin\psi),$$
$$V_{ey} = -V_x \cos\beta \cos\psi - \sin\beta[V_z + V_i(y_b, \psi) + l_{FH}(\Omega_x \sin\psi + \Omega_y \cos\psi)],$$
$$V_{ez} = -V_x \sin\beta \cos\psi + (V_z + V_i(y_b, \psi))\cos\beta + \qquad (3.3)$$
$$+ (y_b + l_{FH}\cos\beta)(\Omega_x \sin\psi + \Omega_y \cos\psi) - y_b \frac{d\beta}{dt}.$$

3.3.4   As it was assumed that blade flapping angles take small values ($\sin\beta$ is close to zero), then the expressions of the air velocity coordinates can be simplified. The third summand in V_{ex} and the second summand in V_{ey} are quite small in respect to their correspondent summands and can be neglected. With these and applying the assumption of small blade flapping angle that $\cos\beta \approx 1$ and $\sin\beta \approx \beta$, the coordinates of the element air velocity can be expressed in a simplified form:

$$V_{ex} = (l_{FH} + y_b)\omega + V_x \sin\psi,$$
$$V_{ey} = -V_x \cos\psi,$$
$$V_{ez} = -V_x\beta\cos\psi + V_z + V_i(y_b, \psi) + (l_{FH} + y_b)(\Omega_x\sin\psi + \Omega_y\cos\psi) - y_b\frac{d\beta}{dt}.$$
$$(3.4)$$

3.3.5 It is common for the blade element rotor theory to normalize the element air velocity components to circumferential velocity of a blade tip, which equals ωR. According to this, the coordinates of the element air velocity can be expressed as:

$$V_{ex} = \omega R\left(\frac{l_{FH} + y_b}{R} + \mu\sin\psi\right),$$
$$V_{ey} = \omega R(-\mu\cos\psi),$$
$$V_{ez} = \omega R\left(-\mu\beta\cos\psi + \lambda(y_b, \psi) + \frac{l_{FH} + y_b}{R}(\bar{\Omega}_x\sin\psi + \bar{\Omega}_y\cos\psi) - \frac{1}{\omega}\frac{y_b}{R}\frac{d\beta}{dt}\right),$$
$$(3.5)$$

where

$$\mu := \frac{V_x}{\omega R}, \quad \lambda(y_b, \psi) := \frac{V_z + V_i(y_b, \psi)}{\omega R}, \quad \bar{\Omega}_x := \frac{\Omega_x}{\omega}, \quad \bar{\Omega}_y := \frac{\Omega_y}{\omega}.$$

The term μ is an advance ratio, which describes forward motion of the rotor system. This term equals zero at hovering and vertical motion of the rotor system; this term is greater than zero at forward motion of the rotor system.

The term $\lambda(y_b, \psi)$ is the inflow ratio of a blade element at position y_b and ψ, which describes the oncoming airflow on the element in the normal direction. The term $\lambda(y_b, \psi)$ includes the induced airflow as well the airflow, which blows a blade element parallelly the rotor axis at vertical motion of the rotor system ($V_z \neq 0$).

The terms $\bar{\Omega}_x$ and $\bar{\Omega}_y$ are normalized lateral and longitudinal angular velocities of the rotor system.

It is useful for further purposes to represent the derivative of blade flapping angle β with respect to time as a derivative with respect to blade azimuthal position ψ on the base of the relationship $\omega = d\psi/dt$:

$$\frac{d\beta}{dt} = \frac{d\psi}{dt}\frac{d\beta}{d\psi} = \omega\frac{d\beta}{d\psi}.$$

The normalized element air velocity coordinates are introduced here with these normalized terms:

$$\bar{V}_{ex} := \frac{V_{ex}}{\omega R} = \frac{l_{FH} + y_b}{R} + \mu\sin\psi,$$
$$\bar{V}_{ey} := \frac{V_{ey}}{\omega R} = -\mu\cos\psi,$$
$$\bar{V}_{ez} := \frac{V_{ez}}{\omega R} = -\mu\beta\cos\psi + \lambda(y_b, \psi) + \frac{l_{FH} + y_b}{R}(\bar{\Omega}_x\sin\psi + \bar{\Omega}_y\cos\psi) - \frac{y_b}{R}\frac{d\beta}{d\psi}.$$
$$(3.6)$$

These normalized coordinates will simplify expressions of the further derivation of rotor forces and moments.

3.4 ATTACK AND SIDESLIP ANGLES OF A BLADE ELEMENT

3.4.1 A reference plane of a blade element is defined here as a plane passing across the element middle point and is parallel to the $x_b z_b$-plane of the blade fixed frame, which is the same as the xz-plane of the blade-hinge frame. Based on the element pitch angle definition (see 2.4.8), the element pitch angle φ_e can be found as the angle between the element chord and a line, which passes the element middle point and is parallel to the tangential x-axis of the blade-hinge frame.

3.4.2 Angle of attack of a blade element (denoted by α_e) is defined as the angle between the blade element chord and a projection of the element air velocity on the element reference plane (fig. 3.2). The element attack angle is positive if the z-coordinate of the element air velocity is positive ($V_{ez} > 0$): this means that the blade moves downward with respect to the rotor system.

Analyzing in the blade-hinge frame, the element attack angle is a sum of the element pitch angle (φ_e) and angle (ε_e) between the element air velocity projection on the reference plane and the x-axis of the blade-hinge frame: $\alpha_e = \varphi_e + \varepsilon_e$. The ε_e can be found from V_{ex} and V_{ez} coordinates of the element air velocity in the blade-hinge frame:

$$\sin \varepsilon_e = \frac{V_{ez}}{\sqrt{V_{ex}^2 + V_{ez}^2}}, \quad \cos \varepsilon_e = \frac{V_{ex}}{\sqrt{V_{ex}^2 + V_{ez}^2}}. \tag{3.7}$$

Taking into account the assumption of $|V_{ez}| \ll V_{ex}$ (see 2.5.5), the ε_e is small enough to be approximated to:

$$\varepsilon_e \approx \sin \varepsilon_e \approx \frac{V_{ez}}{V_{ex}}, \quad \cos \varepsilon_e \approx 1$$

Therefore, the element attack angle can be approximately expressed as:

$$\alpha_e \approx \varphi_e + \frac{V_{ez}}{V_{ex}} \tag{3.8}$$

3.4.3 Sideslip angle of a blade element is defined as angle between the element air velocity and the element reference plane (fig. 3.2). This angle is denoted here as $slip_e$. The element sideslip angle is positive if the V_{ey} coordinate of the element air velocity is positive and directed toward the blade tip. The $slip_e$ can be found in terms of the element air velocity coordinates:

$$\sin(slip_e) = \frac{V_{ey}}{V_e}, \quad \cos(slip_e) = \frac{\sqrt{V_{ex}^2 + V_{ez}^2}}{V_e} \tag{3.9}$$

where V_e is the magnitude of the element air velocity $V_e := |\vec{V}_e|$.

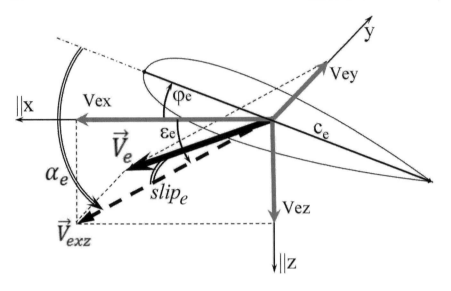

Figure 3.2 Blade element attack and sideslip angles, $\vec{V}_{exz}$ is the projection of element air velocity on element reference plane.

3.5 AERODYNAMIC PROPERTIES OF A BLADE ELEMENT

3.5.1 A blade element is blown by oncoming airflow, which is represented by a vector of the blade air velocity $\vec{V}_e$ in the blade-hinge frame (see 3.3). The oncoming airflow properties are described here by: the volumetric mass density of the surrounded airspace ρ; the airspeed of the blade element $V_e := |\vec{V}_e|$; and the dynamic pressure of the airflow $\rho V_e^2/2$.

3.5.2 A blade element as a wing has an aerodynamic (airfoil) surface, over which pressure distribution is induced at the air blowing. The pressure difference between the upper surface and the underside of the element creates a lift force acting on this element.

Friction of the oncoming airflow with the element airfoil surface creates friction forces, which are distributed over the airfoil surface and are directed tangentially to elementary areas of the airfoil surface, and depend on the airflow velocity. The friction forces over the whole element airfoil surface create profile resistance of the blade element to the motion in the oncoming airflow; the resistance force of the element is called the element profile (parasitic) drag. The element profile drag is directed opposite to the element air velocity.

Superposition of the element lift force and the element profile drag, which act on the blade element, represents the element aerodynamic force $d\vec{R}$.

3.5.3 An aerodynamic force, which acts per an elementary area dA of the element airfoil surface, can be represented by a force $\vec{F}_{dA}$, which is applied to an internal point of the elementary area. Generalizing over the whole airfoil surface of the blade element (A_e), a point can be found, about which the superposition of moments of all aerodynamic forces per elementary areas equals zero. This point is called the element center of pressure and is considered as the point to which the superposition of forces per the elementary areas over the whole airfoil profile is applied without creating any force moment. The position of the pressure center depends on the airfoil profile as well as on air blowing conditions. The position of the element center of pressure does not change along the blade longitudinal axis, is located in the middle between transverse element edges on the element chord, and can change its position only in the element transverse direction along the element chord according to air blow conditions. The aerodynamic force $d\vec{R}$ of the blade element is applied to the center of pressure of this blade element and equals:

$$dR = \oint_{A_e} \vec{F}_{dA}. \tag{3.10}$$

3.5.4   According to the aerodynamic center concept, a stationary point can be found fixed on a blade element chord about which all aerodynamic forces per elementary areas of the element airfoil surface create pitching moment dM_{AD} in the transverse direction of the blade element, which proportionally depends only on the dynamic pressure of the oncoming airflow:

$$dM_{AD} = C_{me} \frac{\rho V_e^2}{2} c_e dS, \tag{3.11}$$

where dS is the planform area of the blade element, c_e is the length of the blade element chord. The pitching moment coefficient C_{me} of the blade element depends only on its airfoil. This stationary point is called the element aerodynamic center and can be described by a radius-vector with coordinates in the blade fixed frame $\vec{p}_{ac}^{(b)}(x_{ac}^{(b)}, y_b, 0)$, where the normal z_b-coordinate equals zero, and y_b is the longitudinal coordinate of the element middle point because the aerodynamic center lays on the element chord.

Within this concept, interaction of a blade element with its oncoming airflow can be represented by:

– the aerodynamic force of the blade element $d\vec{R}$, which is the superposition of all aerodynamic forces per elementary airfoil areas and is applied to the stationary aerodynamic center of the blade element;
– the pitching moment of the blade element dM_{AD}, which acts on the blade element around the blade longitudinal axis and depends only on the oncoming airflow dynamic pressure.

3.5.5   As it was shown above, an aerodynamic force of a blade element ($d\vec{R}$) consists of an element lift force and an element profile drag (fig. 3.3).

A lift force of a blade element with planform area dS is stated here as a vector $d\vec{L}$, which is directed perpendicular to the element air velocity and to the blade longitudinal axis and has the magnitude:

$$dL = C_{Le} \frac{\rho V_e^2}{2} dS. \tag{3.12}$$

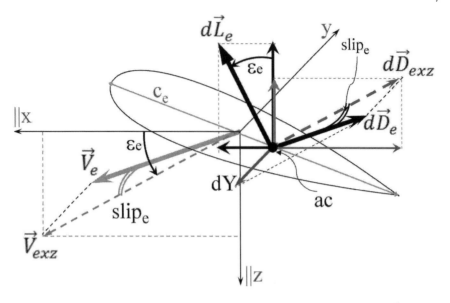

Figure 3.3 Blade element aerodynamic force: *ac* - element aerodynamic center; $\vec{D}_{exz}$ - projection of element drag on element reference plane.

The positive direction of the element lift force is stated toward the upper side of the rotor system. The element lift coefficient C_{Le} determines the property of the blade element to create the lift force. It is assumed here that the element lift coefficient depends only on the element attack angle (α_e) in the oncoming airflow, and this dependence is linearly increasing:

$$C_{Le} = C_{Le}^{\alpha} \alpha_e, \tag{3.13}$$

where C_{Le}^{α} is the element lift coefficient slope, which depends only on the element airfoil. The lift coefficient slope usually takes values in a range of $5.0 - 6.2$. The element lift force can be expressed using this lift coefficient with the defined above element attack angle (3.8):

$$dL = C_{Le}^{\alpha} \left(\varphi_e + \frac{V_{ez}}{V_{ex}} \right) \frac{\rho V_e^2}{2} dS. \tag{3.14}$$

The element profile drag is stated here as a vector $d\vec{D}$ which is directed opposite to the element air velocity vector and has the magnitude:

$$dD = C_{De} \frac{\rho V_e^2}{2} dS, \tag{3.15}$$

where C_{De} is the element profile drag coefficient, which depends only on the element airfoil and is assumed to be constant at any air blowing conditions. The typical value of the profile drag coefficient is about hundredths ($C_{De} \approx 10^{-2}$).

So that, the element aerodynamic force is a vector sum of the element lift force and the element profile drag $d\vec{R} = d\vec{L} + d\vec{D}$ which is applied to the element aerodynamic center.

3.5.6   It is useful to operate with an aerodynamic force of a blade element with coordinates in the blade-hinge frame. The element aerodynamic force $d\vec{R}$ can be decomposed into: the dZ component is the projection of $d\vec{R}$ on the normal z-axis; the dX component is the projection on the tangential x-axis; the dY component is the projection on the longitudinal y-axis. The element lift force and the element drag, which compose the element aerodynamic force, are oriented relative to the element air velocity according to their definitions (fig. 3.3). Orientation of the element aerodynamic force vector $d\vec{R}$ in the blade-hinge frame can be found in terms of the angle ε_e and sideslip angle of the element (see 3.4.3):

$$dX = -dD\cos(slip_e)\cos\varepsilon_e + dL\sin\varepsilon_e,$$
$$dY = -dD\sin(slip_e),$$
$$dZ = -dL\cos\varepsilon_e - dD\cos(slip_e)\sin\varepsilon_e.$$

The element aerodynamic force components can be expressed with the above found sinuses and cosines of angle ε_e (3.7) and sideslip angle (3.9) of the element, as well as using the expression of dL with the determined element attack angle (3.14) and the definition of dD (3.15):

$$dX = -\frac{\rho\,dS}{2}C_{De}V_{ex}V_e + \frac{\rho\,dS}{2}C_{Le}^{\alpha}\left(\varphi_e + \frac{V_{ez}}{V_{ex}}\right)V_{ez}V_e\frac{V_e}{\sqrt{V_{ex}^2 + V_{ez}^2}},$$

$$dY = -\frac{\rho\,dS}{2}C_{De}V_eV_{ey},$$

$$dZ = -\frac{\rho\,dS}{2}C_{Le}^{\alpha}\left(\varphi_e + \frac{V_{ez}}{V_{ex}}\right)V_{ex}V_e\frac{V_e}{\sqrt{V_{ex}^2 + V_{ez}^2}} - \frac{\rho\,dS}{2}C_{De}V_{ez}V_e.$$

As it was assumed (see 2.5.5) that the V_{ex} is the dominant component of the element air velocity ($V_{ex} \gg |V_{ey}|$ and $V_{ex} \gg |V_{ez}|$), then $V_e \approx V_{ex} \approx \sqrt{V_{ex}^2 + V_{ez}^2}$. Based on this assumption, in order to simplify further calculations, it is useful to represent the element airspeed by the V_{ex} coordinate of the air velocity: $V_{ex} \approx V_e$. With these:

$$dX = -\frac{\rho\,dS}{2}C_{De}V_{ex}^2 + \frac{\rho\,dS}{2}C_{Le}^{\alpha}\left(\varphi_e V_{ex}V_{ez} + V_{ez}^2\right),$$

$$dY = -\frac{\rho\,dS}{2}C_{De}V_{ex}V_{ey},$$

$$dZ = -\frac{\rho\,dS}{2}C_{Le}^{\alpha}\left(\varphi_e V_{ex}^2 + V_{ex}V_{ez}\right) - \frac{\rho\,dS}{2}C_{De}V_{ex}V_{ez}.$$

The second summand in dZ represents the projection of the element profile drag on the normal z-axis. Because of small value of the C_{De} in comparison with C_{Le}, this summand is small in respect to another summand ($|C_{De}V_{ex}V_{ez}| \ll |C_{Le}^{\alpha}V_{ex}V_{ez}|$) and can be neglected. With this assumption and with the definition of the element planform area $dS = c_e dy_b$, the components of the element aerodynamic force can be expressed:

$$dX = -\frac{1}{2}\rho c_e C_{De} V_{ex}^2 dy_b + \frac{1}{2}\rho c_e C_{Le}^\alpha V_{ez} \left(\varphi_e V_{ex} + V_{ez}\right) dy_b,$$

$$dY = -\frac{1}{2}\rho c_e C_{De} V_{ex} V_{ey} dy_b, \tag{3.16}$$

$$dZ = -\frac{1}{2}\rho c_e C_{Le}^\alpha \left(\varphi_e V_{ex}^2 + V_{ex} V_{ez}\right) dy_b.$$

3.5.7 There are achieved the components (3.16) of blade element aerodynamic force along the axes of the blade-hinge frame.

The tangential component dX represents the force of element resistance to the circumferential motion due to the rotation around the rotor shaft and acts in the reverse direction to this motion. The first summand $-\frac{1}{2}\rho c_e C_{De} V_{ex}^2 dy_b$ of dX in (3.16) is the element profile drag due to the friction of the airflow with the element airfoil surface; the minus sigh determines the opposite direction of the profile drag to the circumferential motion. The second summand $\frac{1}{2}\rho c_e C_{Le}^\alpha V_{ez}(\varphi_e V_{ex} + V_{ez})dy_b$ of dX is a projection of the element lift force on the tangential x-axis, which is caused by tilt of the element lift force in respect to the normal z-axis on the angle ε_e. The greater is the normal component V_{ez} of the element air velocity at downward element motion relative to its oncoming airflow, the greater is the tilt. If the blade element moves upward relative to the oncoming airflow (air blows the element from above), then the V_{ez} is negative, therefore, the second summand is negative and acts in the opposite direction of the circumferential motion. This force, which is represented by the second summand of dX, is called the element induced drag.

The component dY represents the element profile drag, which appears at longitudinal air blowing of the element, and acts along the longitudinal y-axis. The element longitudinal air blowing appears at forward motion of the rotor system ($V_x > 0$) and depends on the blade azimuthal position: there is no such longitudinal air blowing at $\psi = \pi/2$ and $\psi = 3\pi/2$; the air blowing due to forward motion is directed along the blade longitudinal axis at $\psi = 0$ and is opposite to this axis at $\psi = \pi$.

The normal component dZ mostly represents the element lift force. Since the positive lift force is directed upward with respect to the rotor system, this component is negative as a projection of the element lift force on the normal z-axis, which is directed downward; therefore, there is the minus in front of dZ.

These components of the element aerodynamic force, which is applied to the element aerodynamic center, together with the element pitching moment dM_{AD} (3.11) fully represent the aerodynamic interaction of the element with its oncoming airflow.

3.5.8 It is useful to represent the components of the blade element aerodynamic force in terms of normalized element velocity coordinates introduced in (3.6):

$$dX = \frac{1}{2}\rho c_e(\omega R)^2 \left(-C_{De} \bar{V}_{ex}^2 dy_b + C_{Le}^\alpha \bar{V}_{ez} \left(\varphi_e \bar{V}_{ex} + \bar{V}_{ez}\right) dy_b\right),$$

$$dY = -\frac{1}{2}\rho c_e(\omega R)^2 C_{De} \bar{V}_{ex} \bar{V}_{ey} dy_b, \tag{3.17}$$

$$dZ = -\frac{1}{2}\rho c_e(\omega R)^2 C_{Le}^\alpha \left(\varphi_e \bar{V}_{ex}^2 + \bar{V}_{ex} \bar{V}_{ez}\right) dy_b.$$

Such representation simplifies mathematical derivations of dynamic properties of the rotor system at further analysis.

3.6 BLADE ELEMENTS AS PARTS OF A RIGID BLADE

Here are discussed particularities of a blade element as a part of an absolutely rigid blade. This analysis is important for the generalization of all blade elements in their whole blade. The concept of an absolutely rigid blade body is based on approaches of classical mechanics.

3.6.1 An absolutely rigid blade body consists of a set of whatever big amount (N_p) of (ideal) point particles. Each point particle has a mass and is located within the blade. Position of each point particle is unchanged relative to any other point particle in the absolutely rigid blade. Such unchanged position of every point particle is ensured by mutual interaction of each particle with any other particle of the rigid blade.

A sum of masses of all point particles of the rigid blade equals to mass of the whole blade (mass conservation). All particles are located within the rigid blade volume. Any particle can be affected by some external force or external forces.

Each particle motion obeys the Newton's laws of motion.

According to the Newton's third law, one point particle affects another point particle by an internal force, which is directed along a line joining these two particles, with same magnitude but in opposite direction than an internal force, by which this other particle affects the first particle.

3.6.2 Some j-th point particle within the particle set of the rigid blade ($1 \geq j \geq N_p$) can be described by its mass m_j and its position inside the blade defined by a radius-vector with coordinates in the blade fixed frame $\vec{p}_j^{(b)}(x_{bj}, y_{bj}, z_{bj})$. These coordinates remain unchanged for the absolutely rigid blade. A radius-vector of the j-th point particle position in the blade-hinge frame $\vec{p}_j(x_j, y_{bj}, z_j)$ can be found with the transformation matrix $M^{(b \to h)}$ (2.2) from the blade fixed frame to the blade-hinge frame: $\vec{p}_j = M^{(b \to h)} \vec{p}_j^{(b)}$. The y-coordinate of $\vec{p}_j$ is the same as for $\vec{p}_j^{(b)}$ and remains unchanged because the y-axis of the blade-hinge frame coincides with the y_b-axis of the blade fixed frame. The x_j and z_j coordinates change if the blade pitch angle is changed.

Absolute velocity of the j-th point particle defines the point particle motion relative to the inertial coordinate system and is described by a vector $\vec{v}_j(v_{jx}, v_{jy}, v_{jz})$ with coordinates in the blade-hinge frame. This absolute velocity depends on the blade motion as well as the location of the point particle within the blade. The j-th point particle absolute velocity is determined by absolute velocity (3.1) of the blade point located in the particle position $\vec{p}_j$.

Another k-th point particle ($k \neq j$) affects the j-th point particle by an internal force $\vec{F}_{j,k}^{(inter)}$. All other point particles in the rigid blade affect this j-th point particle by superposition of all internal forces acted on the j-th point particle:

$$\sum_{k \neq j, k=1}^{N_p} \vec{F}_{j,k}^{(inter)}.$$

The j-th point particle can be as well affected by a superposition of external forces. This superposition is denoted by a vector $\vec{F}_{j\Sigma}^{(exter)}$. So that, the j-th point particle is affected by superposition $(\vec{F}_j)$ of its internal and external forces:

$$\vec{F}_j = \sum_{k \neq j, k=1}^{N_p} \vec{F}_{j,k}^{(inter)} + \vec{F}_{j\Sigma}^{(exter)}. \tag{3.18}$$

3.6.3 A point particle is affected by external forces depending on the point particle location in the blade. The following external forces act on a certain point particle.

- The gravity (the particle weight) acts on every point particle due to its mass. The particle weight depends on orientation of the rotor system relative to the global vertical direction, which is determined by the vector of the gravitational acceleration $\vec{g}$. Weight of a j-th point particle with mass m_j acts along the gravitational acceleration vector and equals by magnitude to $m_j|\vec{g}|$. The particle weight is described by coordinates in the blade-hinge frame $m_j\vec{g}^{(h)}(m_j g_x^{(h)}, m_j g_y^{(h)}, m_j g_z^{(h)})$, where $\vec{g}^{(h)}$ is the gravitational acceleration vector with coordinates in the blade-hinge frame.
- If a blade point particle is located on the blade airfoil surface, then the particle is affected by an aerodynamic force. The aerodynamic force of the surface particle is described by an aerodynamic force $\vec{F}_{dA}$ per elementary area dA which represents the surface area related to the point particle (see 3.5.3).
- Point particles, which belong to the blade lugs and contact with the flapping hinge axle and the hinge stopper, are affected by the hinge reaction forces (see 2.4.11). It is assumed that each lug contacts with the hinge axle and the hinge stopper at a single point particle. As was stated that there are two blade lugs on the flapping hinge axle, then there are two contact particles with the hinge axle, which are numbered by a correspondent lug ($i = 1, 2$). The lug contact particles belong to the root part of the blade; therefore, their coordinates in the blade-hinge frame have only tangential coordinates: $\vec{p}_{lug}^{(1)}(x_{lug}^{(1)}, 0, 0)$ and $\vec{p}_{lug}^{(2)}(x_{lug}^{(2)}, 0, 0)$. A i-th lug contact particle ($i = 1, 2$) is affected by a flapping hinge reaction force $\vec{R}_{FH}^{(i)}$ created by the flapping hinge axle and the flapping hinge stopper. The hinge reaction force on the i-th lug contact particle consists of: the axle reaction force along the longitudinal axis $R_{Yaxle}^{(i)}$; the axle reaction force along the normal axis $R_{Zaxle}^{(i)}$; and the hinge slip stopper reaction force $R_{slip}^{(i)}$ directed along the tangential axis. These components compose coordinates of the hinge reaction force on the lug contact particle in the blade-hinge frame $\vec{R}_{FH}^{(i)}(R_{slip}^{(i)}, R_{Yaxle}^{(i)}, R_{Zaxle}^{(i)})$.
- A blade point particle, which locates in the joint of the blade pitch horn with the pitch link from the rotating swashplate, is affected by a reaction force of the pitch link, which was denoted as the blade pitch force $\vec{R}_{pitch}$ (see 2.4.9). It is assumed here that: the pitch horn is a solid part of the blade; and the

horn joins the pitch link at this single point particle. Position of the contact point particle with the pitch link is represented by the radius-vector to this joint with coordinates in the blade fixed frame $\vec{p}_{pitch}(x_{pitch}^{(b)}, y_{pitch}^{(b)}, 0)$. Since the blade rotates together with the rotating swashplate and its pitch link, there is no tangential pulling force. The contact particle is affected by the blade pitch force $\vec{R}_{pitch}$, which has only the normal component in the blade-hinge frame: $\vec{R}_{pitch}(0, 0, R_{pitch})$.

3.6.4 The blade-hinge frame is a non-inertial coordinate system because it rotates relative to the inertial coordinate system. Therefore, the Newton's second law, which defines a differential equation of point particle motion, has an adapted form.

There is a j-th point particle with mass m_j, which moves with absolute velocity $\vec{v}_j$ specified in the blade-hinge frame. A total force $\vec{F}_j$, which is specified in the blade-hinge frame, acts on the point particle. The blade-hinge frame rotates relative to the inertial coordinate system with the angular velocity $\vec{\omega}_h$ with coordinates specified in the blade-hinge frame (2.4). The total force $\vec{F}_j$ acting on the point particle consists of internal forces and external forces (3.18). The superposition of these forces causes the change of the point particle absolute velocity i.e point particle acceleration. Thus, the change of the point particle absolute velocity under these forces is expressed in the blade-hinge (non-inertial rotating) frame:

$$m_j \frac{d\vec{v}_j}{dt} + m_j \vec{\omega}_h \times \vec{v}_j = \sum_{k \neq j, k=1}^{N_p} \vec{F}_{j,k}^{(inter)} + \vec{F}_{j\Sigma}^{(exter)} \tag{3.19}$$

This differential equation describes motion of the j-th point particle. The unchanged position of the point particle within the blade during the blade motion is ensured by the internal forces. This equation is valid for all of the point particles of the blade.

3.6.5 The absolute velocity $\vec{v}_j$ of the j-th point particle is determined by the absolute velocity of the blade point (see 3.2) at the particle position $\vec{p}_j(x_j, y_j, z_j)$ with coordinates in the blade-hinge frame and can be expressed according to (3.1):

$$\vec{v}_j = M^{(r \to h)} \vec{V} + \left(\vec{\omega}^{(h)} + \vec{\Omega}^{(h)} \right) \times \left(\vec{l}_{FH}^{(h)} + \vec{p}_j \right) + \left(-\frac{d\beta}{dt} \vec{e}_x + \frac{d\varphi}{dt} \vec{e}_y \right) \times \vec{p}_j.$$

The left part of the particle motion equation (3.19) is separately analyzed here with the expression of the point particle absolute velocity:

$$m_j \left(\frac{d\vec{v}_j}{dt} + \vec{\omega}_h \times \vec{v}_j \right) = m_j \left[\frac{d\left(M^{(r \to h)} \vec{V} \right)}{dt} + \vec{\omega}_h \times \left(M^{(r \to h)} \vec{V} \right) \right] +$$

$$+ m_j \left\{ \frac{d\left(\left(\vec{\omega}^{(h)} + \vec{\Omega}^{(h)} \right) \times \left(\vec{l}_{FH}^{(h)} + \vec{p}_j \right) + \left(-\frac{d\beta}{dt} \vec{e}_x + \frac{d\varphi}{dt} \vec{e}_y \right) \times \vec{p}_j \right)}{dt} + \right. \tag{3.20}$$

$$\left. + \vec{\omega}_h \times \left(\left(\vec{\omega}^{(h)} + \vec{\Omega}^{(h)} \right) \times \left(\vec{l}_{FH}^{(h)} + \vec{p}_j \right) + \left(-\frac{d\beta}{dt} \vec{e}_x + \frac{d\varphi}{dt} \vec{e}_y \right) \times \vec{p}_j \right) \right\}.$$

The expression in square brackets of the first summand in the right part (3.20) represents a vector $\vec{A}^{(h)}(A_x^{(h)}, A_y^{(h)}, A_z^{(h)})$ of the hub center acceleration relative to the inertial coordinate system with coordinates in the blade-hinge frame:

$$\vec{A}^{(h)} := \frac{d\left(M^{(r\to h)}\vec{V}\right)}{dt} + \vec{\omega}_h \times \left(M^{(r\to h)}\vec{V}\right).$$

The expression in curly brackets of the second summand in the right part of (3.20) represents accelerations of the point particle due to rotational motion of the blade. It must be admitted that the radius-vector $\vec{p}_j$ is not constant in a general case, and its differential depends on rate of blade pitch angle change:

$$\frac{d\vec{p}_j}{dt} = \frac{d\varphi}{dt}\vec{e}_y \times \vec{p}_j.$$

Since the blade-hinge frame rotates relative to the rotor frame due to the shaft rotation ($\vec{\omega}^{(h)}$) and the flapping motion ($-\vec{e}_x d\beta/dt$), then the derivatives of the angular velocities $\vec{\omega}^{(h)}$ and $\vec{\Omega}^{(h)}$ in the blade-hinge frame are expressed in the following way:

$$\frac{d\vec{\omega}^{(h)}}{dt} = \frac{d(M^{(r\to h)}\vec{\omega})}{dt} = M^{(r\to h)}\frac{d\vec{\omega}}{dt} - \left(\vec{\omega}^{(h)} - \frac{d\beta}{dt}\vec{e}_x\right) \times (M^{(r\to h)}\vec{\omega}) =$$

$$= \vec{\varepsilon}_\omega^{(h)} - \left(\vec{\omega}^{(h)} - \frac{d\beta}{dt}\vec{e}_x\right) \times \vec{\omega}^{(h)} = \vec{\varepsilon}_\omega^{(h)} + \frac{d\beta}{dt}\vec{e}_x \times \vec{\omega}^{(h)},$$

$$\frac{d\vec{\Omega}^{(h)}}{dt} = \frac{d(M^{(r\to h)}\vec{\Omega})}{dt} = M^{(r\to h)}\frac{d\vec{\Omega}}{dt} - \left(\vec{\omega}^{(h)} - \frac{d\beta}{dt}\vec{e}_x\right) \times (M^{(r\to h)}\vec{\Omega}) =$$

$$= \vec{\varepsilon}_\Omega^{(h)} - \left(\vec{\omega}^{(h)} - \frac{d\beta}{dt}\vec{e}_x\right) \times \vec{\Omega}^{(h)},$$

where $\vec{\varepsilon}_\omega^{(h)} := M^{(r\to h)}(d\vec{\omega}/dt)$ is the angular acceleration of the rotor shaft rotation with coordinates in the blade-hinge frame, $\vec{\varepsilon}_\Omega^{(h)} := M^{(r\to h)}(d\vec{\Omega}/dt)$ is the angular acceleration of the rotor system rotation with coordinates in the blade-hinge frame.

3.6.6   The particle motion equation (3.19) can be rewritten with the left part (3.20) after expanding of $\vec{\omega}_h$ according to (2.3), differentiation, and sorting of correspondent terms:

$$m_j\left(-\frac{d^2\beta}{dt^2}\vec{e}_x \times \vec{p}_i + \frac{d^2\varphi}{dt^2}\vec{e}_y \times \vec{p}_j + (\vec{\varepsilon}_\omega^{(h)} + \vec{\varepsilon}_\Omega^{(h)}) \times (\vec{l}_{FH}^{(h)} + \vec{p}_j)\right) +$$

$$+ m_j\left[\vec{\omega}^{(h)} \times \left(\vec{\omega}^{(h)} \times (\vec{l}_{FH}^{(h)} + \vec{p}_j)\right) + \vec{\Omega}^{(h)} \times \left(\vec{\Omega}^{(h)} \times (\vec{l}_{FH}^{(h)} + \vec{p}_j)\right) +$$

$$+ \left(\frac{d\beta}{dt}\right)^2 \vec{e}_x \times (\vec{e}_x \times \vec{p}_j) + \left(\frac{d\varphi}{dt}\right)^2 \vec{e}_y \times (\vec{e}_y \times \vec{p}_j)\right] +$$

$$+ m_j\left\{2\vec{\Omega}^{(h)} \times \left(\omega^{(h)} \times (\vec{l}_{FH}^{(h)} + \vec{p}_j) - \frac{d\beta}{dt}\vec{e}_x \times \vec{p}_j + \frac{d\varphi}{dt}\vec{e}_y \times \vec{p}_j\right) +$$

$$+ 2\vec{\omega}^{(h)} \times \left(-\frac{d\beta}{dt}\vec{e}_x \times \vec{p}_j + \frac{d\varphi}{dt}\vec{e}_y \times \vec{p}_j\right) - 2\frac{d\beta}{dt}\frac{d\varphi}{dt}\vec{e}_x \times (\vec{e}_y \times \vec{p}_j)\right\} + \qquad (3.21)$$

$$+ m_j\vec{A}^{(h)} = \sum_{k\neq j, k=1}^{N_p} \vec{F}_{j,k}^{(inter)} + \vec{F}_{j\Sigma}^{(exter)}.$$

Analyzing in the non-inertial blade-hinge frame, the total force, which is applied to the point particle, causes the particle motion with: the rotational acceleration (in parentheses); the centripetal acceleration (in square brackets); the Coriolis acceleration (in curly brackets); as well as with translational acceleration ($\vec{A}^{(h)}$). According to the classical mechanics approach for non-inertial coordinate systems, there are introduced (fictitious) inertial forces: centrifugal forces are represented by vectors opposite to those, which are with square brackets; Coriolis forces are represented by vectors opposite to those, which are with curly brackets; Euler forces are represented by vectors opposite to those, which are with round parentheses, and appear at accelerating rotations; the term $-m_j\vec{A}^{(h)}$ is a translational inertial force, which acted on the point particle at accelerated motion of the whole rotor system.

Following centrifugal forces act on the point particle:

- The term $-m_j\vec{\omega}^{(h)} \times \left(\vec{\omega}^{(h)} \times (\vec{l}_{FH}^{(h)} + \vec{p}_j) \right)$ represents a centrifugal force due to rotation around the rotor shaft, which is directed perpendicularly from the rotor axis.
- The term $-m_j\vec{\Omega}^{(h)} \times \left(\vec{\Omega}^{(h)} \times (\vec{l}_{FH}^{(h)} + \vec{p}_j) \right)$ represents a centrifugal force due to rotation of the whole rotor system, which is directed perpendicularly from direction of $\vec{\Omega}$ laying in the rotor plane.
- The term $-m_j(d\beta/dt)^2\vec{e}_x \times (\vec{e}_x \times \vec{p}_j)$ represents a centrifugal force due to the blade flapping motion, which is directed perpendicularly from the flapping axis (the longitudinal x-axis).
- The term $-m_j(d\varphi/dt)^2\vec{e}_y \times (\vec{e}_y \times \vec{p}_j)$ represents a centrifugal force due to rotation around the blade feathering hinge at blade pitch angle change, which is directed perpendicularly from the longitudinal y-axis.

The mutual rotations of the blade in the inertial system cause a set of Coriolis forces acting on the blade point particle:

- The term $-2m_j\vec{\Omega}^{(h)} \times (\omega^{(h)} \times (\vec{l}_{FH}^{(h)} + \vec{p}_j))$ is a Coriolis force due to the rotation around the rotor shaft accompanied by the rotation of the rotor system and is directed along the z_r-axis of the rotor frame; this force is tilted to the normal z-axis at flapping angle and is quite close to the axis due to small value of the flapping angle. If the angular velocity $\vec{\Omega}^{(h)}$ is directed along azimuthal position ψ_Ω, then: this Coriolis force acts downward (with respect to the rotor system) on a blade particle at the azimuthal position ψ_Ω; this force acts upward at $\pi + \psi_\Omega$; this force does not affects a point particle at $\psi_\Omega + \pi/2$ and $\psi_\Omega - \pi/2$.
- The term $2m_j\vec{\Omega}^{(h)} \times ((d\beta/dt)\vec{e}_x \times \vec{p}_j)$ is a Coriolis force due to blade flapping motion accompanied by the rotation of the rotor system; the vector is directed at flapping angle to the rotor plane and perpendicular to the angular velocity of the rotor system rotation $\vec{\Omega}$.
- The term $-2m_j\vec{\Omega}^{(h)} \times ((d\varphi/dt)\vec{e}_y \times \vec{p}_j)$ is a Coriolis force due to rotation around the feathering hinge (blade pitch angle change) accompanied by the rotation of the rotor system and is perpendicular to the angular velocity of the rotor system rotation.

- The term $2m_j\vec{\omega}^{(h)} \times ((d\beta/dt)\vec{e}_x \times \vec{p}_j)$ is a Coriolis force due to blade flapping motion accompanied by the rotation around the rotor shaft and is directed close to the tangential x-axis.
- The term $-2m_j\vec{\omega}^{(h)} \times ((d\varphi/dt)\vec{e}_y \times \vec{p}_j)$ is a Coriolis force due to the rotation around the blade feathering hinge accompanied by blade flapping motion and is perpendicular to the rotor axis.
- The term $2m_j(d\beta/dt)(d\varphi/dt)\vec{e}_x \times (\vec{e}_y \times \vec{p}_j)$ is a Coriolis force due to rotation around the blade feathering hinge accompanied by blade flapping motion and directed along the blade longitudinal y-axis.

3.6.7   Motion of the whole blade can be described by a system of the motion equations of all blade point particles with the condition of the third Newton's law, which specifies a relationship between internal forces:

$$
\begin{cases}
m_1 \dfrac{d\vec{v}_1}{dt} + m_1 \vec{\omega}_h \times \vec{v}_1 = \displaystyle\sum_{k=2}^{N_p} \vec{F}_{1,k}^{(inter)} + \vec{F}_{1\Sigma}^{(exter)} \\[2mm]
\dots \\[1mm]
m_j \dfrac{d\vec{v}_j}{dt} + m_j \vec{\omega}_h \times \vec{v}_j = \displaystyle\sum_{k\neq j, k=1}^{N_p} \vec{F}_{j,k}^{(inter)} + \vec{F}_{j\Sigma}^{(exter)} \\[2mm]
\dots \\[1mm]
m_{Np} \dfrac{d\vec{v}_{Np}}{dt} + m_{Np} \vec{\omega}_h \times \vec{v}_{Np} = \displaystyle\sum_{k=1}^{N_p-1} \vec{F}_{Np,k}^{(inter)} + \vec{F}_{Np\Sigma}^{(exter)} \\[2mm]
\dots \\[1mm]
\vec{F}_{j,k>j}^{(inter)} = -\vec{F}_{k,j}^{(inter)} \\[1mm]
\dots
\end{cases} \quad ,
$$

where $1 \geq j \geq N_p$, $1 \geq k \geq N_p$.

The direct solving of this equation system is avoided here. The solution of the equation system is considered in terms of properties of the blade elements. In the frame of the blade element rotor theory, the equation system of blade motion is subdivided into equation systems, each of which refers to a single blade element.

3.6.8   According to the blade element approach mentioned above, a blade is transversely divided into blade elements. Each blade element is described by position of its middle point y_b along the blade longitudinal axis and by its element length dy_b. The blade element contains a subset n_p of blade point particles ($n_p \in N_p$), which are located within the volume of this blade element. Motion of the blade element is described by a system of motion equations of point particles, which belong to the blade element:

$$
\begin{cases}
m_i \dfrac{d\vec{v}_i}{dt} + m_i \vec{\omega}_h \times \vec{v}_i = \displaystyle\sum_{k\neq i, k=1}^{N_p} \vec{F}_{i,k}^{(inter)} + \vec{F}_{i\Sigma}^{(exter)} \\[2mm]
\dots \\[1mm]
\vec{F}_{i,j>i}^{(inter)} = -\vec{F}_{j,i}^{(inter)} \\[1mm]
\dots
\end{cases} \quad ,
$$

where i and j are indexes of particles, which belong to the blade element ($i \in n_p$, $j \in n_p$); k is an index of particles, which belong to the whole blade, not just the element ($k \in N_p$). The superposition of internal forces $\sum_{k \neq i, k=1}^{N_p} \vec{F}_{i,k}^{(inter)}$, which act on a i-th point particle, can be represented by a sum of: internal forces created by other particles in the blade element $\sum_{j \in n_p, j \neq i} \vec{F}_{i,k}^{(inter)}$; and internal forces created by particles outside the blade element $\sum_{l \in N_p \backslash n_p} \vec{F}_{i,l}^{(inter)}$:

$$\begin{cases} m_i \dfrac{d\vec{v}_i}{dt} + m_i \vec{\omega}_h \times \vec{v}_i = \sum_{j \in n_p, j \neq i} \vec{F}_{i,j}^{(inter)} + \sum_{l \in N_p \backslash n_p} \vec{F}_{i,l}^{(inter)} + \vec{F}_{i\Sigma}^{(exter)} \\ \dots \\ \vec{F}_{i,j>i}^{(inter)} = -\vec{F}_{j,i}^{(inter)} \\ \dots \end{cases} \quad , \quad (3.22)$$

where $N_p \backslash n_p$ denotes the set of blade point particles, which do not belong to the elements, l is an index of a particle, which belongs to the blade but does not belong to the blade element ($l \in N_p \backslash n_p$).

There are two ways of generalization of the equation system for particles, which belong to a single element. The first one is based on the definition of a center of mass and describes the element displacement; the second one describes the element rotation. Further, these two ways are particularly discussed.

3.7   EQUATION OF MOTION OF BLADE ELEMENT CENTER OF MASS

3.7.1   One differential equation can be achieved by adding all differential equations of the system (3.22) for particles belonging to one blade element:

$$\begin{aligned} \frac{d \sum_{i \in n_p} m_i \vec{v}_i}{dt} + \vec{\omega}_h \times \sum_{i \in n_p} m_i \vec{v}_i = \\ = \sum_{\substack{i \in n_p \\ j \neq i}} \sum_{j \in n_p} \vec{F}_{i,j}^{(inter)} + \sum_{i \in n_p} \sum_{l \in N_p \backslash n_p} \vec{F}_{i,l}^{(inter)} + \sum_{i \in n_p} \vec{F}_{i\Sigma}^{(exter)}. \end{aligned} \quad (3.23)$$

According to the concept of a center of mass for an absolutely rigid body, a rigid blade element has a unique point, which is stationary with respect to this blade element, is described by a radius-vector $\vec{p}_e$, and about which a sum of weighted position vectors of all element point particles constantly equals zero:

$$\sum_{i \in n_p} m_i (\vec{p}_i - \vec{p}_e) = 0, \quad (3.24)$$

where $m_i (\vec{p}_i - \vec{p}_e)$ is a weighted position vector of a i-th point particle about this stationary point $\vec{p}_e$. This stationary point is the element center of mass. As stated above, the element center of mass is located on the blade longitudinal axis and coincides with the middle point of the element (see 3.1.1). The total momentum of all of the element point particles can be described as a momentum of an imagined point particle with the mass of the whole element, which is located in the element center of mass:

$$\sum_{i \in n_p} m_i \vec{v}_i = \frac{\hat{d} \sum_{i \in n_p} m_i \vec{p}_i}{dt} = \frac{\hat{d} \sum_{i \in n_p} m_i (\vec{p}_i - \vec{p}_e)}{dt} + \frac{\hat{d} \sum_{i \in n_p} m_i \vec{p}_e}{dt} = \frac{\hat{d} \vec{p}_e}{dt} \sum_{i \in n_p} m_i,$$

or

$$\sum_{i \in n_p} m_i \vec{v}_i = dm \vec{v}_e,$$

where $\vec{v}_e = \hat{d}\vec{p}_e/dt$ is the blade element absolute velocity (see 3.2.3), dm is the mass of the whole blade element, which equals to the sum of masses of all element particles according to the mass conservation: $dm = \sum_{i \in n_p} m_i$. The differentials $\hat{d}\vec{p}_i$ and $\hat{d}\vec{p}_e$ represent here infinitely small displacements measured in the inertial coordinate system with coordinates in the blade-hinge frame.

3.7.2   The first summand in the right part of the equation (3.23) represents the mutual interaction between point particles within the element. According to the Newton's third law, mutually interacting particles (i-th and j-th) affect each other by the forces, which are equal by magnitude but act in opposite directions ($\vec{F}_{i,j} = -\vec{F}_{j,i}$). The nested sum can be reordered in the following way, which justifies that it equals zero:

$$\sum_{i \in n_p} \sum_{\substack{j \in n_p \\ j \neq i}} \vec{F}_{i,k}^{(inter)} = \sum_{i \in n_p} \sum_{\substack{j \in n_p \\ j > i}} \left(\vec{F}_{i,j} + \vec{F}_{j,i} \right) = 0.$$

The second summand in the right part of the equation (3.23) represents the interaction between particles inside the element with particles outside the element. This nested sum can be interpreted as a force difference $d\vec{F}_e^{(inter)}$ between interactions of the element with the rest blade parts from both sides of the element. This force difference is defined as difference between the interaction of the element with blade particles $\vec{F}_{eFar}^{(inter)}$, which are out of the element far from the blade flapping hinge, and the interaction of the element with blade particles $\vec{F}_{eClose}^{(inter)}$, which are out of the element closer to the flapping hinge:

$$\sum_{i \in n_p} \sum_{l \in N_p \setminus n_p} \vec{F}_{i,l}^{(inter)} = \vec{F}_{eFar}^{(inter)} - \vec{F}_{eClose}^{(inter)} = d\vec{F}_e^{(inter)}.$$

This means that the internal forces create stress inside the element in all directions. The total internal force is described with coordinates in the blade-hinge frame $d\vec{F}_e^{(inter)}(dF_{ex}^{(inter)}, dF_{ey}^{(inter)}, dF_{ez}^{(inter)})$, where $dF_{ex}^{(inter)}$ is the element stress force in the tangential direction, $dF_{ey}^{(inter)}$ is the element tension force, and $dF_{ez}^{(inter)}$ is the element stress force in the normal direction.

3.7.3   The third summand in the right part of the equation (3.23) represents a superposition of all external forces, which act on particles in the element. As mentioned in 3.6.3, there are four kinds of external forces, which depend on the element position in its blade.

Each particle of the element, which is located on the element airfoil surface A_e, is affected by an aerodynamic force $\vec{F}_{dA}$ per an elementary area dA, which refers to this particle. A sum of the aerodynamic forces per elementary areas over the element airfoil surfaces represents the element aerodynamic force $d\vec{R}$ (3.10).

All element particles are affected by the gravity. Each element particle with mass m_i has a weight $\vec{g}^{(h)} m_i$. All particles of an element compose the weight of the element $\vec{g}^{(h)} dm$, where dm is the mass of the blade element.

The element aerodynamic force and the element wight are present for any blade element and mostly represent the total external force of the element:

$$\sum_{i \in n_p} \vec{F}_{i\Sigma}^{(exter)} = \vec{g}^{(h)} \sum_{i \in n_p} m_i + \oint_{A_e} \vec{F}_{dA} = dm\vec{g}^{(h)} + d\vec{R}.$$

The root element ($y_b = 0$) contains two particles, which contact with the flapping hinge axle. Each of these two particles is affected by the correspondent flapping hinge reaction force described in 2.4.11 and 3.6.3: $\vec{R}_{FH}^{(1)}(R_{slip}^{(1)}, R_{Yaxle}^{(1)}, R_{Zaxle}^{(1)})$ and $\vec{R}_{FH}^{(2)}(R_{slip}^{(2)}, R_{Yaxle}^{(2)}, R_{Zaxle}^{(2)})$. Therefore, the total flapping hinge reaction force $\vec{R}_{FH} = \vec{R}_{FH}^{(1)} + \vec{R}_{FH}^{(2)}$ additionally acts on the root element:

$$\sum_{i \in n_p} \vec{F}_{i\Sigma}^{(exter)}\bigg|_{y_b=0} = dm\vec{g}^{(h)} + d\vec{R} + \vec{R}_{FH}^{(1)} + \vec{R}_{FH}^{(2)} = dm\vec{g}^{(h)} + d\vec{R} + \vec{R}_{FH}.$$

The blade element, which contains the blade pitch horn joined with the pitch link from the rotating swashplate, is affected by the blade pitch force $\vec{R}_{pitch}(0, 0, R_{pitch})$:

$$\sum_{i \in n_p} \vec{F}_{i\Sigma}^{(exter)}\bigg|_{y_b=y_{pitch}} = dm\vec{g}^{(h)} + d\vec{R} + \vec{R}_{pitch}.$$

3.7.4   According to the all specified above terms, the equation (3.23) is rewritten in the form of the motion equation of the center of mass of the blade element with the mass dm under external and internal forces:

$$dm\left(\frac{d\vec{v}_e}{dt} + \vec{\omega}_h \times \vec{v}_e\right) = d\vec{F}_e^{(inter)} + d\vec{F}_e^{(exter)}, \tag{3.25}$$

where

$$d\vec{F}_e^{(exter)} := dm\vec{g}^{(h)} + d\vec{R} + \vec{F}_{react}(y_b),$$

is the superposition of the external forces. The term $\vec{F}_{react}(y_b)$ specifies: the flapping hinge reaction force for the root blade element $\vec{F}_{react}(0) = \vec{R}_{FH}$; the blade pitch force $\vec{F}_{react}(y_{pitch}) = \vec{R}_{pitch}$ acting on the element at the position, where the pitch horn joins the pitch link; the term equals zero for all other elements.

The left part of the element motion equation (3.25) corresponds to the left part of the motion equation (3.19) and (3.21) for an imagined point particle, which is located at the blade element center of mass $\vec{p}_e(0, y_b, 0)$, has the mass of the blade element dm, and has the element absolute velocity $\vec{v}_e$. The left part of the element motion equation can be rewritten in an extended form with this interpretation according to the left part of the point particle motion equation (3.21). Since the element mass center lays on the longitudinal axis, the position of the blade element center of mass can be represented in vector form $\vec{p}_e = y_b \vec{e}_y$, and the terms, which are related to rotation around the feathering hinge, are reduced:

$$dm\left(-y_b \frac{d^2 \beta}{dt^2} \vec{e}_z + (\vec{\varepsilon}_\omega^{(h)} + \vec{\varepsilon}_\Omega^{(h)}) \times (\vec{l}_{FH}^{(h)} + y_b \vec{e}_y) \right) +$$

$$+ dm\left[\vec{\omega}^{(h)} \times \left(\vec{\omega}^{(h)} \times (\vec{l}_{FH}^{(h)} + y_b \vec{e}_y) \right) + \vec{\Omega}^{(h)} \times \left(\vec{\Omega}^{(h)} \times (\vec{l}_{FH}^{(h)} + y_b \vec{e}_y) \right) - \right.$$

$$\left. - y_b \left(\frac{d\beta}{dt} \right)^2 \vec{e}_y \right] + \tag{3.26}$$

$$+ dm\left\{ 2\vec{\Omega}^{(h)} \times \left(\omega^{(h)} \times (\vec{l}_{FH}^{(h)} + y_b \vec{e}_y) - y_b \frac{d\beta}{dt} \vec{e}_z \right) - 2y_b \frac{d\beta}{dt} \vec{\omega}^{(h)} \times \vec{e}_z \right\} =$$

$$= d\vec{F}_e^{(inter)} + d\vec{F}_e^{(exter)} - m_j \vec{A}^{(h)}.$$

This equation describes motion of the blade element center of mass under external and internal forces as well as the inertial forces. The internal forces ensure the motion of the element as a part of the rigid blade. The element external forces participate in motion of the whole blade. The inertial forces due to the rotation around the feathering hinge do not participate in the motion of the element center of mass.

This equation of motion of the element center of mass will be further used to find the flapping hinge reaction force. Force, with which the blade acts on its flapping hinge, will be found based on the flapping hinge reaction force.

3.8   EQUATION OF BLADE ELEMENT ROTATION

3.8.1   One differential equation, which describes motion of a blade element caused by rotations of its blade, can be achieved by adding all differential equations of the element particles system (3.22), each equation of which is previously modified by a cross-product on a radius-vector of a correspondent point particle:

$$\sum_{i \in n_p} m_i \left(\vec{p}_i \times \frac{d\vec{v}_i}{dt} \right) + \sum_{i \in n_p} m_i \left(\vec{p}_i \times (\vec{\omega}_h \times \vec{v}_i) \right) =$$

$$= \sum_{i \in n_p} \sum_{\substack{j \in n_p \\ j \neq i}} \vec{p}_i \times \vec{F}_{i,j}^{(inter)} + \sum_{i \in n_p} \sum_{l \in N_p \setminus n_p} \vec{p}_i \times \vec{F}_{i,l}^{(inter)} + \sum_{i \in n_p} \vec{p}_i \times \vec{F}_{i\Sigma}^{(exter)}. \tag{3.27}$$

3.8.2   The first term in the right part of the equation (3.27) represents the sum of moments around the flapping hinge center created by the internal forces between the point particles within the blade element. This nested sum can be reordered with respect to a sum of the force moments of two mutually interacting particles (i-th and j-th):

$$\sum_{\substack{i\in n_p}} \sum_{\substack{j\in n_p \\ j\neq i}} \vec{p}_i \times \vec{F}_{i,k}^{(inter)} = \sum_{\substack{i\in n_p}} \sum_{\substack{j\in n_p \\ j>i}} \left(\vec{p}_i \times \vec{F}_{i,j} + \vec{p}_j \times \vec{F}_{j,i} \right) = 0.$$

According to the Newton's third law, two forces ($\vec{F}_{i,j}$ and $\vec{F}_{j,i}$) of two interacting particles have equal magnitude, act along a line, which joins these particles, but in opposite directions. Sum moment of such two forces equals zero because these forces, which are opposite and equal by magnitude, act along a same line: $\vec{p}_i \times \vec{F}_{i,j} + \vec{p}_j \times \vec{F}_{j,i} = 0$. Therefore the whole nested sum equal zero.

The second term in the right part of the equation (3.27) represents a sum of moments of forces between blade particles inside the blade element with blade particles outside the element. The nested sum can be interpreted as force moment difference, which affects the element due to different internal interactions of the element with rest blade parts from both sides of this element. This moment difference describes the internal torque created during the blade motion and acts on the element in all directions. The internal torque is the difference between a total force moment of interactions of the element with blade particles far from the flapping hinge ($\vec{M}_{eFar}^{(inter)}$) and a total force moment of interactions of the element with blade particles close to the flapping hinge ($\vec{M}_{eClose}^{(inter)}$):

$$\sum_{i\in n_p} \sum_{l\in N_p \backslash n_p} \vec{p}_i \times \vec{F}_{i,l}^{(inter)} = \vec{M}_{eFar}^{(inter)} - \vec{M}_{eClose}^{(inter)} = d\vec{M}_e^{(inter)}.$$

The internal torque is described by a vector with coordinates in the blade-hinge frame $d\vec{M}_e^{(inter)}(dM_{ex}^{(inter)}, dM_{ey}^{(inter)}, dM_{ez}^{(inter)})$, where $dM_{ex}^{(inter)}$ is the bending moment in the tangential direction, $dM_{ey}^{(inter)}$ is the twisting stress, and $dM_{ez}^{(inter)}$ is the bending moment in the normal direction.

3.8.3   The third summand in the right part of the equation (3.27) represents a superposition of moments about the flapping hinge center, which are created by external forces acting on all particles in the element.

According to the defined above element mass center, that a superposition of weighted position vectors about the element center of mass equals zero (3.24), the total element gravity force is applied to the element center of mass and creates the force moment about the flapping hinge center:

$$\sum_{i\in n_p} m_i \vec{p}_i \times \vec{g}^{(h)} = dm\vec{p}_e \times \vec{g}^{(h)}.$$

A particle of the element, which lays on the element airfoil surface A_e, is affected by an aerodynamic force $\vec{F}_{dA}$ per an elementary area dA referred to this particle. The superposition $d\vec{R}$ of the aerodynamic forces per elementary areas over the element airfoil surface is applied to the element aerodynamic center $\vec{p}_{ac}^{(b)}(x_{ac}^{(b)}, y_b, 0)$, which is located on the blade element chord and creates the force moment $(M^{(b\to h)}\vec{p}_{ac}^{(b)}) \times d\vec{R}$ about the flapping hinge center (see 3.5.4). Besides this moment, the aerodynamic forces create the element aerodynamic pitch moment dM_{AD} directed around the y-longitudinal axis (3.11).

Element aerodynamic moments and a moment of an element wight are present for any blade element and mostly represent the total external force moments:

$$\sum_{i \in n_p} \vec{p}_i \times \vec{F}_{i\Sigma}^{(exter)} = dm\vec{p}_e \times \vec{g}^{(h)} + (M^{(b \to h)} \vec{p}_{ac}^{(b)}) \times d\vec{R} + \vec{e}_y dM_{AD}.$$

Two particles of the root element ($y_b = 0$), which belong to the blade lugs, contact with the blade flapping hinge and are affected by the axle reaction forces and the stopper reaction forces as described in 2.4.11 and 3.6.3. Due to the separate location of the lugs on the flapping hinge axle, the flapping hinge reaction forces creates a moment around the normal z-axis of the blade-hinge frame acting on the root element:

$$M_{FH} = x_{lug}^{(1)} R_{Yaxle}^{(1)} + x_{lug}^{(2)} R_{Yaxle}^{(2)},$$

where $x_{lug}^{(1)}$ and $x_{lug}^{(2)}$ are tangential coordinates of the contacts of the lag with the axle. The moment M_{FH} was defined as the flapping hinge reaction moment in 2.4.11.

The blade element, which contains the joint of the blade pitch horn with the pitch link described by the radius-vector $\vec{p}_{pitch}^{(b)}(x_{pitch}^{(b)}, y_{pitch}^{(b)}, 0)$ in the blade fixed frame, is affected by the blade pitch force $\vec{R}_{pitch}(0, 0, R_{pitch})$ with coordinate in the blade-hinge frame as described in 2.4.9 and 3.6.3. This force creates the moment $\vec{M}_{pitch}$ with coordinates in the blade-hinge frame:

$$\vec{M}_{pitch} = \left(M^{(b \to h)} \vec{p}_{pitch}^{(b)} \right) \times \vec{R}_{pitch}.$$

For further derivation simplification, assuming small values of the element pitch angle (see 2.5.6), the position of the pitch horn joint with the pitch link in the blade-hinge frame $\vec{p}_{pitch}$ approximately has the same coordinates as the position of this joint in the blade fixed frame $\vec{p}_{pitch}^{(b)}$:

$$\vec{p}_{pitch} = M^{(b \to h)} \vec{p}_{pitch}^{(b)} \approx \vec{p}_{pitch}^{(b)}.$$

The equation (3.27) is rewritten in the following way with all of the specified above terms:

$$\sum_{i \in n_p} \left(m_i \vec{p}_i \times \frac{d\vec{v}_i}{dt} + m_i \vec{p}_i \times (\vec{\omega}_h \times \vec{v}_i) \right) = d\vec{M}_e^{(inter)} + d\vec{M}_e^{(exter)}, \tag{3.28}$$

where

$$d\vec{M}_e^{(exter)} := dm\vec{p}_e \times \vec{g}^{(h)} + \vec{p}_{ac} \times d\vec{R} + \vec{e}_y dM_{AD} + \vec{M}_{react}(y_b)$$

is the moment of external forces, which acts on the element. The term $\vec{M}_{react}(y_b)$ is a moment of the external reaction forces and specifies: the flapping hinge reaction moment acting on the root element $\vec{M}_{react}(0) = \vec{e}_z M_{FH}$; the blade pitch force moment $\vec{M}_{react}(y_{pitch}) = \vec{M}_{pitch}$ acting on the blade element joined to the pitch link; the term equals zero for all other elements.

3.8.4   Before analysis of the left part of the equation (3.28), here is analyzed the left part of the point particle motion equation (3.21) for some i-th particle multiplied using a cross product on a radius vector $\vec{p}_i$ of the i-th particle:

$$
m_i \vec{p}_i \times \frac{d\vec{v}_i}{dt} + m_i \vec{p}_i \times (\vec{\omega}_h \times \vec{v}_i) =
$$

$$
= m_i \vec{p}_i \times \left[\left(-\frac{d^2\beta}{dt^2} \vec{e}_x + \frac{d^2\varphi}{dt^2} \vec{e}_y + \vec{\varepsilon}_\omega^{(h)} + \vec{\varepsilon}_\Omega^{(h)} \right) \times \vec{p}_i \right] +
$$

$$
+ m_i \vec{p}_i \times \left[(\vec{\varepsilon}_\omega^{(h)} + \vec{\varepsilon}_\Omega^{(h)}) \times \vec{l}_{FH}^{(h)} \right] +
$$

$$
+ m_i \left[\vec{p}_i \times \left[\vec{\omega}^{(h)} \times (\vec{\omega}^{(h)} \times \vec{p}_i) \right] + \vec{p}_i \times \left[\vec{\Omega}^{(h)} \times (\vec{\Omega}^{(h)} \times \vec{p}_i) \right] +
$$

$$
+ \vec{p}_i \times \left[\vec{\omega}^{(h)} \times (\vec{\omega}^{(h)} \times \vec{l}_{FH}^{(h)}) \right] + \vec{p}_i \times \left[\vec{\Omega}^{(h)} \times (\vec{\Omega}^{(h)} \times \vec{l}_{FH}^{(h)}) \right] +
$$

$$
+ \left(\frac{d\beta}{dt} \right)^2 \vec{p}_i \times [\vec{e}_x \times (\vec{e}_x \times \vec{p}_i)] + \left(\frac{d\varphi}{dt} \right)^2 \vec{p}_i \times [\vec{e}_y \times (\vec{e}_y \times \vec{p}_i)] \right] + \tag{3.29}
$$

$$
+ m_i \left\{ 2\vec{p}_i \times \left[\vec{\Omega}^{(h)} \times \left(\left(\omega^{(h)} - \frac{d\beta}{dt} \vec{e}_x + \frac{d\varphi}{dt} \vec{e}_y \right) \times \vec{p}_i \right) \right] +
$$

$$
+ 2\vec{p}_i \times \left[\vec{\omega}^{(h)} \times \left(\left(-\frac{d\beta}{dt} \vec{e}_x + \frac{d\varphi}{dt} \vec{e}_y \right) \times \vec{p}_i \right) \right] +
$$

$$
+ 2\vec{p}_i \times \left[\vec{\Omega}^{(h)} \times \left(\omega^{(h)} \times \vec{l}_{FH}^{(h)} \right) \right] - 2\frac{d\beta}{dt}\frac{d\varphi}{dt} \vec{p}_i \times [\vec{e}_x \times (\vec{e}_y \times \vec{p}_i)] \right\} +
$$

$$
+ m_i \vec{p}_i \times \vec{A}^{(h)}.
$$

Essentially, this part represents the change rate of the particle angular moment relative to the bade-hinge frame under internal and external forces acting on the particle. The huge expression is going to be represented in terms of mechanical inertia measures of a rotating point particle relative to the blade-hinge frame. These introduced measures also simplify the equation to a certain extent.

There is the i-th particle with mass m_i and the radius vector $\vec{p}_i(x_i, y_i, z_i)$ of its position with coordinates in the blade-hinge frame, and there is an arbitrary vector $\vec{w}$ with coordinates in the same frame, then an expression $m_i \vec{p}_i \times (\vec{w} \times \vec{p}_i)$ can be rewritten according to the triple product (Lagrange's formula) in the following way:

$$
m_i \vec{p}_i \times (\vec{w} \times \vec{p}_i) = m_i \vec{w}(\vec{p}_i \cdot \vec{p}_i) - m_i \vec{p}_i(\vec{w} \cdot \vec{p}_i)
$$

or in a matrix form:

$$
m_i \vec{p}_i \times (\vec{w} \times \vec{p}_i) =
$$

$$
= m_i \left((x_i^2 + y_i^2 + z_i^2) \begin{bmatrix} 1 & 0 & 0 \\ 0 & 1 & 0 \\ 0 & 0 & 1 \end{bmatrix} - \begin{bmatrix} x_i^2 & x_i y_i & x_i z_i \\ y_i x_i & y_i^2 & y_i z_i \\ z_i x_i & z_i y_i & z_i^2 \end{bmatrix} \right) \vec{w},
$$

This expression is useful to write with a tensor I_i:

$$
m_i \vec{p} \times (\vec{w} \times \vec{p}) = I_i \vec{w} \tag{3.30}
$$

where

$$I_i := \begin{bmatrix} m_i y_i^2 + m_i z_i^2 & -m_i x_i y_i & -m_i x_i z_i \\ -m_i y_i x_i & m_i x_i^2 + m_i z_i^2 & -m_i y_i z_i \\ -m_i z_i x_i & -m_i z_i y_i & m_i x_i^2 + m_i y_i^2 \end{bmatrix}.$$

The tensor I_i represents moments of inertia of the i-th point particle in the blade-hinge frame. The diagonal elements of the tensor represent the moment of inertia of the i-th particle around the correspondent axis of the blade-hinge frame: $(m_i y_i^2 + m_i z_i^2)$ is the particle inertia moment around the tangential x-axis, $(m_i x_i^2 + m_i z_i^2)$ – around the longitudinal y-axis, $(m_i x_i^2 + m_i y_i^2)$ – around the normal z-axis. The non-diagonal elements are cross products of inertia, and describe non-symmetrical particle location relative to the axes of the blade-hinge frame.

The expression $-m_i \vec{p} \times (\vec{w} \times \vec{p})$ is interpreted as a moment of an Euler force acting on the i-th point particle at accelerated rotation with an angular acceleration described by a vector $\vec{w}$ in the blade-hinge frame. For example, an Euler force moment at accelerated rotation around the rotor shaft with angular acceleration $\vec{\varepsilon}_\omega^{(h)}$ can be expressed in a short form:

$$- m_i \vec{p} \times (\vec{\varepsilon}_\omega^{(h)} \times \vec{p}) = -I_i \vec{\varepsilon}_\omega^{(h)}.$$

Consider an expression $m_i \vec{p}_i \times \left[\vec{W} \times \left(\vec{W} \times \vec{p}_i \right) \right]$, which is similar to the expression of a centrifugal force moment acting on the i-th particle at rotation of the blade-hinge frame with an arbitrary angular velocity $\vec{W}$. The expression can be rewritten with the Jacobi identity of the external cross product:

$$m_i \vec{p}_i \times \left[\vec{W} \times \left(\vec{W} \times \vec{p}_i \right) \right] = m_i \vec{W} \times \left[\vec{p}_i \times \left(\vec{W} \times \vec{p}_i \right) \right] - m_i \left[\vec{W} \times \vec{p}_i \right] \times \left[\vec{p}_i \times \vec{W} \right] =$$
$$= m_i \vec{W} \times \left[\vec{p}_i \times \left(\vec{W} \times \vec{p}_i \right) \right].$$

The achieved expression is useful to represent in terms of to the defined above identity (3.30):

$$m_i \vec{p}_i \times \left[\vec{W} \times \left(\vec{W} \times \vec{p}_i \right) \right] = \vec{W} \times I_i \vec{W}. \tag{3.31}$$

For example, the moment of the centrifugal force acting on the i-th point particle at rotation around the rotor shaft with the angular velocity $\vec{\omega}^{(h)}$ is written in the short form with the tensor of inertia moment:

$$- m_i \vec{p}_i \times \left[\vec{\omega}^{(h)} \times \left(\vec{\omega}^{(h)} \times \vec{p}_i \right) \right] = -\vec{\omega}^{(h)} \times I_i \vec{\omega}^{(h)}.$$

Consider an expression $m_i \vec{p}_i \times \left[\vec{W} \times \left(\vec{U} \times \vec{p}_i \right) \right]$, which corresponds to a moment of a Coriolis force due to mutual two different rotations with arbitrary angular velocities $\vec{W}$ and $\vec{U}$. The internal cross product can be rewritten with the triple product

(Lagrange's formula):

$$m_i \vec{p}_i \times \left[\vec{W} \times \left(\vec{U} \times \vec{p}_i \right) \right] = m_i \vec{p}_i \times \left[\vec{U} \left(\vec{W} \cdot \vec{p}_i \right) - \vec{p}_i \left(\vec{U} \cdot \vec{W} \right) \right] =$$

$$= m_i \left(\vec{W} \cdot \vec{p}_i \right) \vec{p}_i \times \vec{U} - \left(m_i \vec{U} \cdot \vec{W} \right) \vec{p}_i \times \vec{p}_i = -\vec{U} \times \left[m_i \vec{p}_i \left(\vec{W} \cdot \vec{p}_i \right) \right].$$

The term in the square brackets is useful to represent with a tensor I_{Ci}:

$$m_i \vec{p}_i \left(\vec{W} \cdot \vec{p}_i \right) = I_{Ci} \vec{W},$$

where

$$I_{Ci} := \begin{bmatrix} m_i x_i^2 & m_i x_i y_i & m_i x_i z_i \\ m_i y_i x_i & m_i y_i^2 & m_i y_i z_i \\ m_i z_i x_i & m_i z_i y_i & m_i z_i^2 \end{bmatrix}.$$

This tensor I_{Ci} is used for expressions of moments caused by Coriolis forces acting on the i-th point particle. This tensor differs from the inertia moment tensor I_i by the diagonal elements; the non-diagonal elements are opposite by the signs. The expression with this tensor has the following view:

$$m_i \vec{p}_i \times \left[\vec{W} \times \left(\vec{U} \times \vec{p}_i \right) \right] = -\vec{U} \times I_{Ci} \vec{W}. \tag{3.32}$$

For example, the term, which relates to the moment of the Coriolis force due to rotation around the rotor shaft ($\vec{\omega}^{(h)}$) accompanied by rotor system rotation ($\vec{\Omega}^{(h)}$), can be written in a short form with the tensor I_{Ci}:

$$m_i \vec{p}_i \times \left[\vec{\Omega}^{(h)} \times \left(\vec{\omega}^{(h)} \times \vec{p}_i \right) \right] = -\vec{\omega}^{(h)} \times I_{Ci} \vec{\Omega}^{(h)}.$$

The expression of the left part of the i-th point particle motion equation, which is multiplied using the cross product on the point particle radius vector $\vec{p}_i$ as shown above (3.29), takes the simplified form with the introduced tensor of the point particle moment of inertia and with the tensor for Coriolis moments:

$$m_i \vec{p}_i \times \frac{d\vec{v}_i}{dt} + m_i \vec{p}_i \times \left(\vec{\omega}_h \times \vec{v}_i \right) =$$

$$= I_i \left(-\frac{d^2\beta}{dt^2} \vec{e}_x + \frac{d^2\varphi}{dt^2} \vec{e}_y + \vec{\varepsilon}_\omega^{(h)} + \vec{\varepsilon}_\Omega^{(h)} \right) + m_i \vec{p}_i \times \left(\left(\vec{\varepsilon}_\omega^{(h)} + \vec{\varepsilon}_\Omega^{(h)} \right) \times \vec{l}_{FH}^{(h)} \right) +$$

$$+ \left[\vec{\omega}^{(h)} \times I_i \vec{\omega}^{(h)} + \vec{\Omega}^{(h)} \times I_i \vec{\Omega}^{(h)} + \left(\frac{d\beta}{dt} \right)^2 \vec{e}_x \times I_i \vec{e}_x + \left(\frac{d\varphi}{dt} \right)^2 \vec{e}_y \times I_i \vec{e}_y + \right.$$

$$\left. + \vec{p}_i \times \left(\vec{\omega}^{(h)} \times \left(\vec{\omega}^{(h)} \times \vec{l}_{FH}^{(h)} \right) \right) + \vec{p}_i \times \left(\vec{\Omega}^{(h)} \times \left(\vec{\Omega}^{(h)} \times \vec{l}_{FH}^{(h)} \right) \right) \right] -$$

$$- \left\{ 2 \left(\omega^{(h)} - \frac{d\beta}{dt} \vec{e}_x + \frac{d\varphi}{dt} \vec{e}_y \right) \times I_{Ci} \vec{\Omega}^{(h)} + 2 \left(-\frac{d\beta}{dt} \vec{e}_x + \frac{d\varphi}{dt} \vec{e}_y \right) \times I_{Ci} \vec{\omega}^{(h)} - \right.$$

$$\left. - 2 \frac{d\beta}{dt} \frac{d\varphi}{dt} \vec{e}_y \times I_{Ci} \vec{e}_x - 2 \vec{p}_i \times \left[\vec{\Omega}^{(h)} \times \left(\omega^{(h)} \times \vec{l}_{FH}^{(h)} \right) \right] \right\} + m_j \vec{p}_i \times \vec{A}^{(h)}$$

$$\tag{3.33}$$

3.8.5   The left part of the equation (3.28) represents the sum of left parts (3.33) of motion equations for all point particles of the element, which are previously modified by a cross-product on a radius-vector of a correspondent point particle. A tensor of the element moment of inertia relative to the blade-hinge frame dI is introduced here in order to represent the equation in terms of the element inertial properties and to simplify the equation representation:

$$
dI := \begin{bmatrix}
\sum\limits_{i \in n_p} m_i y_i^2 + m_i z_i^2 & -\sum\limits_{i \in n_p} m_i x_i y_i & -\sum\limits_{i \in n_p} m_i x_i z_i \\
-\sum\limits_{i \in n_p} m_i x_i y_i & \sum\limits_{i \in n_p} m_i x_i^2 + m_i z_i^2 & -\sum\limits_{i \in n_p} m_i y_i z_i \\
-\sum\limits_{i \in n_p} m_i x_i z_i & -\sum\limits_{i \in n_p} m_i y_i z_i & \sum\limits_{i \in n_p} m_i x_i^2 + m_i y_i^2
\end{bmatrix}.
$$

This tensor of element inertia moment equals a sum of tensors of inertia moment of all point particles within the element relative to the blade-hinge frame:

$$
dI = \sum_{i \in n_p} I_i
$$

The equation (3.28) can be rewritten with this tensor and with transferring the moments of centrifugal, Coriolis and translational inertial forces to the right part:

$$
dI \left(-\frac{d^2\beta}{dt^2}\vec{e}_x + \frac{d^2\varphi}{dt^2}\vec{e}_y + \vec{\varepsilon}_\omega^{(h)} + \vec{\varepsilon}_\Omega^{(h)} \right) + dm\vec{p}_e \times \left((\vec{\varepsilon}_\omega^{(h)} + \vec{\varepsilon}_\Omega^{(h)}) \times \vec{l}_{FH}^{(h)} \right) =
$$

$$
= dM_e^{(inter)} + dM_e^{(exter)} - dm\vec{p}_e \times \vec{A}^{(h)} -
$$

$$
- \left[\vec{\omega}^{(h)} \times dI\vec{\omega}^{(h)} + \vec{\Omega}^{(h)} \times dI\vec{\Omega}^{(h)} + \left(\frac{d\beta}{dt}\right)^2 \vec{e}_x \times dI\vec{e}_x + \left(\frac{d\varphi}{dt}\right)^2 \vec{e}_y \times dI\vec{e}_y + \right.
$$

$$
\left. + dm\vec{p}_e \times \left(\vec{\omega}^{(h)} \times (\vec{\omega}^{(h)} \times \vec{l}_{FH}^{(h)}) \right) + dm\vec{p}_e \times \left(\vec{\Omega}^{(h)} \times (\vec{\Omega}^{(h)} \times \vec{l}_{FH}^{(h)}) \right) \right] +
$$

$$
+ \left\{ 2\left(-\frac{d\beta}{dt}\vec{e}_x + \frac{d\varphi}{dr}\vec{e}_y \right) \times dI_C(\vec{\omega}^{(h)} + \vec{\Omega}^{(h)}) + 2\vec{\omega}^{(h)} \times dI_C\vec{\Omega}^{(h)} - \right.
$$

$$
\left. - 2\frac{d\beta}{dt}\frac{d\varphi}{dt}\vec{e}_y \times dI_C\vec{e}_x - 2dm\vec{p}_e \times \left(\vec{\Omega}^{(h)} \times \left(\vec{\omega}^{(h)} \times \vec{l}_{FH} \right) \right) \right\}.
$$

$$
(3.34)
$$

The achieved equation describes the angular rotations of a blade element together with its blade under external forces and is called the equation of the blade element rotation.

The left part of the blade element rotation equation represents angular momentum change of the element in the blade-hinge frame. The first summand represents the element angular momentum change relative to the flapping hinge center; the second summand appears due to the flapping hinge offset at the accelerated rotation around the rotor shaft and the accelerated rotation of the rotor system.

The right part represents a sum of moments of the internal and external forces as well as the moments of the inertial forces, which affect the blade element. The term $-dm\vec{p}_e \times \vec{A}^{(h)}$ affects the element at accelerated translational motion of the rotor system. The terms in square brackets represent moments created by centrifugal forces. The terms in curly brackets represent moments created by Coriolis forces. A tensor dI_C for Coriolis moments is introduced here in order to simplify the equation representation; this tensor is a sum of all tensors for Coriolis moments of all particles within the element; this tensor differs from the moment of inertia tensor (dI) by diagonal elements and by signs:

$$dI_C := \sum_{i \in n_p} I_{Ci} = \begin{bmatrix} \sum_{i \in n_p} m_i x_i^2 & \sum_{i \in n_p} m_i x_i y_i & \sum_{i \in n_p} m_i x_i z_i \\ \sum_{i \in n_p} m_i x_i y_i & \sum_{i \in n_p} m_i y_i^2 & \sum_{i \in n_p} m_i y_i z_i \\ \sum_{i \in n_p} m_i x_i z_i & \sum_{i \in n_p} m_i y_i z_i & \sum_{i \in n_p} m_i z_i^2 \end{bmatrix}.$$

3.8.6   An element inertia tensor dI in the blade-hinge frame is not constant and depends on pitch angle of the element. It is useful to operate with an inertia tensor $dI^{(b)}$ of a blade element in the blade fixed frame, relative to which the element is fixed; therefore, elements of this tensor are constant. This tensor $dI^{(b)}$ does not represent the central principal moments of a blade element inertia around the central principal inertia axes, which pass the element center of mass.

As stated above, one central principle inertia axis of an element is aligned to the blade longitudinal y_b-axis, another coincides with the element chord, third is perpendicular to the previous two. If the element chord would be parallel to the blade transverse x-axis, then the element inertia tensor in the blade fixed frame $dI^{(b)}$ would be diagonal. However, the element chord may not be parallel to the blade transverse axis due to blade twisting ($\varphi_{tw}(y_b) \neq 0$); in this case, the inertia tensor of the blade element in the fixed frame may have non-zero elements in the first row and the third column same as in the third row and the first column. The element inertia tensor relative to the blade fixed frame can be expressed in a general case:

$$dI^{(b)} = \begin{bmatrix} dI_{xx}^{(b)} & 0 & dI_{xz}^{(b)} \\ 0 & dI_{yy}^{(b)} & 0 \\ dI_{xz}^{(b)} & 0 & dI_{zz}^{(b)} \end{bmatrix},$$

where:
$$dI_{xx}^{(b)} = \sum_{i \in n_p} m_i(y_{bi}^2 + z_{bi}^2), \quad dI_{yy}^{(b)} = \sum_{i \in n_p} m_i(x_{bi}^2 + z_{bi}^2), \quad dI_{zz}^{(b)} = \sum_{i \in n_p} m_i(x_{bi}^2 + y_{bi}^2)$$
are the element moments of inertia around the axes of the blade fixed frame;
$$dI_{xz}^{(b)} = -\sum_{i \in n_p} m_i x_{bi} z_{bi}$$
is a product of inertia and depends on the element twisting angle, this term equals zero at $\varphi_{tw}(y_b) = 0$. Since the blade has the high wing aspect ratio, then the coordinates y_{bi} mostly dominate in comparison to other coordinates x_{bi} and z_{bi}: $y_{bi} \gg x_{bi}$, $y_{bi} \gg z_{bi}$. Therefore, the term $dI_{xz}^{(b)}$ is small in comparison to other terms and can be assumed to equal zero: $dI_{xz}^{(b)} \approx 0$. The term $dI_{yy}^{(b)}$ is small as well in comparison

to $dI_{xx}^{(b)}$ and $dI_{zz}^{(b)}$; however, this term will be considered in the analysis of element rotation around the longitudinal y-axis.

The inertia tensor of the blade element in the blade-hinge frame dI can be found on the base of the inertia tensor in the blade fixed frame $dI^{(b)}$ with the known transformation matrix from the blade fixed frame to the blade-hinge frame $M^{(b \to h)}$ (2.2) and with specified blade pitch angle φ:

$$dI = M^{(b \to h)} dI^{(b)} \left(M^{(b \to h)} \right)^T,$$

$$dI = \begin{bmatrix} dI_{xx}^{(b)} \cos^2 \varphi + dI_{zz}^{(b)} \sin^2 \varphi & 0 & -(dI_{xx}^{(b)} - dI_{zz}^{(b)}) \cos \varphi \sin \varphi \\ 0 & dI_{yy}^{(b)} & 0 \\ -(dI_{xx}^{(b)} - dI_{zz}^{(b)}) \cos \varphi \sin \varphi & 0 & dI_{zz}^{(b)} \cos^2 \varphi + dI_{xx}^{(b)} \sin^2 \varphi \end{bmatrix}.$$

The blade pitch angle takes small values to neglect $\sin \varphi$ as well as $\sin^2 \varphi$; the approximation $\cos \varphi \approx 1$ is accepted with this condition. The term $dI_{xx}^{(b)} - dI_{zz}^{(b)}$ is negligibly small for the blade with the high wing aspect ratio. With this explanation, it is reasonable to accept for further derivation simplification that the element inertia tensor in the blade-hinge frame can be approximated to the diagonal tensor with the blade element moments of inertia around the axes of the blade fixed frame:

$$dI \approx dI^{(b)} \approx \begin{bmatrix} dI_{xx}^{(b)} & 0 & 0 \\ 0 & dI_{yy}^{(b)} & 0 \\ 0 & 0 & dI_{zz}^{(b)} \end{bmatrix}. \tag{3.35}$$

With this diagonal tensor of the element moment of inertia dI, the moments due to centrifugal forces caused by the flapping motion and the rotation around the feathering hinge equal zero:

$$\left(\frac{d\beta}{dt} \right)^2 \vec{e}_x \times dI \vec{e}_x = 0, \quad \left(\frac{d\varphi}{dt} \right)^2 \vec{e}_y \times dI \vec{e}_y = 0.$$

The same consideration can be done concerning the tensor for Coriolis moments:

$$dI_C \approx \begin{bmatrix} dI_{Cxx}^{(b)} & 0 & 0 \\ 0 & dI_{Cyy}^{(b)} & 0 \\ 0 & 0 & dI_{Czz}^{(b)} \end{bmatrix}. \tag{3.36}$$

where $dI_{Cxx}^{(b)} = \sum\limits_{i \in n_p} m_i x_{bi}^2$, $dI_{Cyy}^{(b)} = \sum\limits_{i \in n_p} m_i y_{bi}^2$, $dI_{Czz}^{(b)} = \sum\limits_{i \in n_p} m_i z_{bi}^2$.

3.8.7 It is useful for further analysis to represent the blade element rotation equation (3.34) as a set of equations, each of which refers to a correspondent coordinate of the blade-hinge frame.

An equation for the tangential x-coordinate describes element motion around the blade flapping hinge and has the full form:

$$
\begin{aligned}
-dI_{xx}^{(b)}\frac{d^2\beta}{dt^2} =& -\left(dI_{xx}^{(b)}+dmy_bl_{FH}\cos\beta\right)\left(\frac{d\Omega_x}{dt}\sin\psi+\frac{d\Omega_y}{dt}\cos\psi\right)+ \\
& +\left(\left(dI_{zz}^{(b)}-dI_{yy}^{(b)}\right)\cos\beta+dmy_bl_{FH}\right)\sin\beta\,\omega^2- \\
& -\left[\left(dI_{zz}^{(b)}-dI_{yy}^{(b)}\right)\cos\beta\sin\beta\left(\Omega_x\cos\psi-\Omega_y\sin\psi\right)^2\right. \\
& \left. -dmy_bl_{FH}\sin\beta\left(\Omega_x\sin\psi+\Omega_y\cos\psi\right)^2\right]- \\
& -2\left(dI_{Cyy}^{(b)}\cos^2\beta+dI_{Czz}^{(b)}\sin^2\beta+dmy_bl_{FH}\cos\beta\right)\omega(\Omega_x\cos\psi-\Omega_y\sin\psi)- \\
& -2dI_{Czz}^{(b)}\frac{d\varphi}{dt}\omega\cos\beta-2dI_{Czz}^{(b)}\frac{d\varphi}{dt}\sin\beta\left(\Omega_x\cos\psi-\Omega_y\sin\psi\right)+ \\
& +dM_{ex}^{(inter)}+dM_{ex}^{(exter)}-dmy_bA_z^{(h)}
\end{aligned}
$$

The term dmy_bl_{FH} is used as a summand for an element inertia moment about the flapping hinge center in order to get the element inertia moment about the rotor hub center; this summand depends on the flapping hinge offset l_{FH}.

The first summand in the right part is a x-coordinate of an inertial force moment acting on the element due to the accelerated rotation of the rotor system. This term is quite small and would be neglected; however, this term remains in order to analyze an influence of this accelerated rotation on blade flapping motion.

A centrifugal force due to the rotation around the rotor shaft affects the blade element and is always directed perpendicularly from the rotor axis. If the blade lays in the rotor plane, then this centrifugal force is directed along the blade longitudinal y-axis and does not create any moment. If the blade has non-zero flapping angle, then this centrifugal force creates a moment around the flapping hinge, which is proportional to the sinus of the flapping angle and directed opposite to blade flapping. The moment of the centrifugal force due to the rotation around the rotor shaft always tries to eliminate the flapping angle. This moment is represented by the second summand in the right part of the equation. Using the assumption of small flapping angle $\sin\beta\approx\beta$ and $\cos\beta\approx1$, this summand can be simplified to $(dI_{zz}^{(b)}-dI_{yy}^{(b)}+dmy_bl_{FH})\omega^2\beta$.

The third summand (within square brackets) in the right part represents a x-coordinate of a moment of a centrifugal force acting on the element due to the whole rotor system rotation. The first term in square brackets describes a moment component of this centrifugal force caused by a component of the angular velocity, which is directed along a projection of the blade on the rotor plane: $-\Omega_x\cos\psi+\Omega_y\sin\psi$. The second term in the square brackets describes a moment component of this centrifugal force caused by an angular velocity component aligned parallelly to the tangential x-axis: $\Omega_x\sin\psi+\Omega_y\cos\psi$. Both terms have orders of magnitudes of $\approx(dI_{zz}^{(b)}-dI_{yy}^{(b)})\beta(\Omega_x^2+\Omega_y^2)$ and $\approx dmy_bl_{FH}\beta(\Omega_x^2+\Omega_y^2)$. Taking into account that the rotor system rotation is much slower than the rotation around the rotor shaft $|\Omega_x| \ll \omega$ and $|\Omega_y| \ll \omega$ (see 2.5.3) and that β takes small values, the total third summand is negligibly small in comparison to the second summand in the right part and can be neglected.

The fourth summand in the right part represents a x-component of a moment of a Coriolis force due to the mutual rotation of the rotor system with the rotation around the rotor shaft. Due to the high wing aspect ratio of the blade ($dI_{Cyy}^{(b)} \gg dI_{Czz}^{(b)}$) and due to small flapping angle ($\sin\beta \ll \cos\beta$ and $\cos\beta \approx 1$), the multiplier in parenthesis can be simplified to $dI_{Cyy}^{(b)} + dmy_b l_{FH}$.

The fifth summand in the right part represents a x-component of a moment of a Coriolis force due to the mutual rotation around the feathering hinge with rotation around the rotor shaft. The swashplate provides element pitch angle in certain blade azimuthal position ψ according to blade flapping angle β and swashplate position, which is determined by collective pitch θ_0, lateral cyclic pitch θ_1 and longitudinal cyclic pitch θ_2, as specified in (2.7). The element pitch angle is changed during rotation around the shaft; the rate of the element pitch angle change can be expressed as:

$$\frac{d\varphi}{dt} = \theta_1 \omega \sin\psi - \theta_2 \omega \cos\psi - k\frac{d\beta}{dt}.$$

where k is the flapping compensation coefficient. With this and with $\cos\beta \approx 1$, the fifth summand can be expressed as:

$$2dI_{Czz}^{(b)} \omega \frac{d\varphi}{dt} = 2dI_{Czz}^{(b)} \left(\theta_1 \omega^2 \sin\psi - \theta_2 \omega^2 \cos\psi - k\omega\frac{d\beta}{dt} \right).$$

The sixth summand in the right part represents a projection on the x-axis at angle $\pi/2 - \beta$ of a moment of a Coriolis force due to the mutual rotation around the feathering hinge with the rotor system rotation. Because of small value of $dI_{Czz}^{(b)}$ and due to $|\Omega_x| \ll \omega$, $|\Omega_y| \ll \omega$, as well as small flapping angle sinus, this term can be neglected.

The term $-dm\vec{y}_b A_z^{(h)}$ represents an inertial force due to the accelerated translational motion of the whole rotor system acting on the element. As it was assumed in 2.5.2, the rotorcraft acceleration is small enough to neglect this inertial force.

The term $dM_{ex}^{(exter)}$ represents the total moment of the external forces (3.28) acting on the element around the tangential x-axis. This moment has following components:

$$dM_{ex}^{(exter)} = y_b dZ + dmy_b g_z^{(h)} + M_{x(react)}(y_b).$$

The x-coordinate of the moment of the element aerodynamic force is caused by the normal component dZ of the element aerodynamic force (see 3.5.7) with an arm y_b from the flapping hinge to the element position. The x-coordinate of the moment of the element weight is represented by $dmy_b g_z^{(h)}$. The element weight is negligibly small in comparison with the element aerodynamic forces as assumed in 2.5.1; therefore, the element weight moment can be neglected. The x-coordinate of the moment of the reaction forces $M_{x(react)}(y_b)$ defined in (3.28) is represented only by a moment of the blade pitch force R_{pitch} with the arm $y_{pitch}^{(b)}$, which acts only on the blade element containing the joint of the pitch horn with the pitch link, $M_{x(react)}(y_{pitch}^{(b)}) = y_{pitch}^{(b)}R_{pitch}$; the x-coordinate of the moment of reaction forces equals zero for all other elements $(M_{x(react)}(y_b \neq y_{pitch}^{(b)}) = 0)$.

With all of the specified above, the equation of element rotation around the x-axis can be expressed in a simplified form:

$$
\begin{aligned}
dI_{xx}^{(b)}\frac{d^2\beta}{dt^2} &= \left(dI_{xx}^{(b)} + dmy_b l_{FH}\right)\left(\frac{d\Omega_x}{dt}\sin\psi + \frac{d\Omega_y}{dt}\cos\psi\right) - \\
&\quad - \left(dI_{zz}^{(b)} - dI_{yy}^{(b)} + dmy_b l_{FH}\right)\omega^2\beta + \\
&\quad + 2\left(dI_{Cyy}^{(b)} + dmy_b l_{FH}\right)\omega(\Omega_x\cos\psi - \Omega_y\sin\psi) - \\
&\quad - 2dI_{Czz}^{(b)}k\omega\frac{d\beta}{dt} + 2dI_{Czz}^{(b)}\omega^2(\theta_1\sin\psi - \theta_2\cos\psi) - \\
&\quad - dM_{ex}^{(inter)} - y_b dZ - M_{x(react)}(y_b)
\end{aligned}
\tag{3.37}
$$

3.8.8   The element rotation equation (3.34) for the longitudinal y-coordinate describes a blade element rotation around the blade longitudinal axis and has the full form:

$$
\begin{aligned}
dI_{yy}^{(b)}\frac{d^2\varphi}{dt^2} &= dI_{yy}^{(b)}\cos\beta\left(\frac{d\Omega_x}{dt}\cos\psi - \frac{d\Omega_y}{dt}\sin\psi\right) - dI_{yy}^{(b)}\sin\beta\frac{d\omega}{dt} + \\
&\quad + (dI_{xx}^{(b)} - dI_{zz}^{(b)})\sin\beta(\Omega_x\sin\psi + \Omega_y\cos\psi)(\Omega_x\cos\psi - \Omega_y\sin\psi) - \\
&\quad - 2dI_{Cxx}^{(b)}\omega\cos\beta(\Omega_x\sin\psi + \Omega_y\cos\psi) - 2dI_{Czz}^{(b)}\frac{d\beta}{dt}\omega\cos\beta - \\
&\quad - 2dI_{Czz}^{(b)}\frac{d\beta}{dt}\sin\beta(\Omega_x\cos\psi - \Omega_y\sin\psi) + dM_{ey}^{(inter)} + dM_{ey}^{(exter)}
\end{aligned}
$$

$$\tag{3.38}$$

The left part represents the blade element inertia to the accelerated rotation around the feathering hinge. Since the element pitch angle is posed according to swashplate position and tilt (θ_1 and θ_2), as well as to the blade flapping angle β, as specified in (2.7), then the angular acceleration around the feathering hinge is determined by these parameters:

$$
\frac{d^2\varphi}{dt^2} = \theta_1\omega^2\cos\psi + \theta_2\omega^2\sin\psi + \frac{d\omega}{dt}(\theta_1\sin\psi - \theta_2\cos\psi) - k\frac{d^2\beta}{dt^2}.
$$

Assuming, that the rotor shaft angular velocity is kept almost constant ($\omega \approx const$), the expression can be approximated:

$$
\frac{d^2\varphi}{dt^2} = \omega^2(\theta_1\cos\psi + \theta_2\sin\psi) - k\frac{d^2\beta}{dt^2}.
$$

The first summand in the right part of the y-rotation equation (3.38) represents the y-coordinate of the inertial force moment due to the accelerated rotation of the rotor system.

The second summand in the right part is a projection on the longitudinal axis at angle $\pi/2 - \beta$ of the inertial force moment due to the accelerated rotation around the rotor shaft. The rotor shaft rotation is usually is kept constant, therefore, this angular acceleration is small. Besides this, this projection depends on sinus of the blade flapping angle; since the flapping angle is small, then the projection is small as well. Based on these, this summand can be neglected.

The third summand is a projection on the longitudinal y-axis at angle $\pi/2 - \beta$ of a moment of a centrifugal force due to rotation of the rotor system. This summand is neglected because of small angular velocities of the rotor system rotation ($|\Omega_x| \ll \omega$ and $|\Omega_y| \ll \omega$) as well as small value of $dI_{xx}^{(b)} - dI_{zz}^{(b)}$ and small value of $\sin\beta$.

The fourth summand is a moment of a Coriolis force due to mutual rotation of the rotor system with the rotation around the rotor shaft. The fifth summand is a moment of a Coriolis force due to mutual rotation of the rotor shaft with the flapping motion. Both summands have the same order of magnitude as the left part of the equation and are not neglected.

The sixth summand represents a projection on the longitudinal axis at angle $\pi/2 - \beta$ of a moment of a Coriolis force due to mutual rotation of the rotor system with the flapping motion. This term is two orders of magnitude smaller than the left part of the equation due to small values of the angular velocities ($|\Omega_x| \ll \omega$ and $|\Omega_y| \ll \omega$) and small value of the flapping angle ($\sin\beta \approx 0$) and can be neglected.

The term $dM_{ey}^{(exter)}$ represents a projection on the longitudinal y-axis of the moments of the external forces (3.28) and has the following components:

$$dM_{ey}^{(exter)} = -x_{ac}dZ + dM_{AD} + M_{y(react)}(y_b).$$

The y-coordinate of the moment of the element aerodynamic force is caused by the normal component dZ of the element aerodynamic force (see 3.5.4) with an arm x_{ac} from the longitudinal axis to the element aerodynamic center and by the element aerodynamic pitching moment dM_{AD} (3.11). Since the element center of mass lays on the longitudinal y-axis, the element weight does not create the moment around this axis. The y-component of the moment of the reaction forces $M_{y(react)}(y_b)$ is represented only by the moment of the blade pitch force R_{pitch} with the arm $x_{pitch}^{(b)}$, which acts on the blade element attached to the pitch horn $M_{y(react)}(y_{pitch}^{(b)}) = -x_{pitch}^{(b)}R_{pitch}$; this component equals zero for all other elements $M_{y(react)}(y_b \neq y_{pitch}) = 0$.

With all of the stated above and with the approximation $\cos\beta \approx 1$, the equation of element rotation around the longitudinal y-axis is expressed as a differential equation with respect to β:

$$dI_{yy}^{(b)}k\frac{d^2\beta}{dt^2} = -dI_{yy}^{(b)}\left(\frac{d\Omega_x}{dt}\cos\psi - \frac{d\Omega_y}{dt}\sin\psi\right) +$$
$$+ dI_{yy}^{(b)}\omega^2(\theta_1\cos\psi + \theta_2\sin\psi) + 2dI_{Cxx}^{(b)}\omega(\Omega_x\sin\psi + \Omega_y\cos\psi) +$$
$$+ 2dI_{Czz}^{(b)}\frac{d\beta}{dt}\omega - dM_{ey}^{(inter)} + x_{ac}dZ - dM_{AD} - M_{y(react)}(y_b).$$

$$(3.39)$$

3.8.9   The element rotation equation (3.34) for the normal z-coordinate describes a blade element rotation around direction perpendicular to the flapping hinge axis and to the longitudinal blade y-axis and has the full form:

$$-(dI_{zz}^{(b)}\cos\beta + dmy_b l_{FH})\frac{d\omega}{dt} = dI_{zz}^{(b)}\sin\beta\left(\frac{d\Omega_x}{dt}\cos\psi - \frac{d\Omega_y}{dt}\sin\psi\right) -$$

$$-\left((dI_{xx}^{(b)} - dI_{yy}^{(b)})\cos\beta + dmy_b l_{FH}\right)(\Omega_x\sin\psi + \Omega_y\cos\psi)(\Omega_x\cos\psi - \Omega_y\sin\psi) -$$

$$-2dI_{Cyy}^{(b)}\omega\sin\beta\frac{d\beta}{dt} + 2dI_{Cyy}^{(b)}\cos\beta\frac{d\beta}{dt}(\Omega_x\cos\psi - \Omega_y\sin\psi) -$$

$$-2dI_{Cxx}^{(b)}\left[\frac{d\varphi}{dt}(\Omega_x\sin\psi + \Omega_y\cos\psi) + \omega\sin\beta(\Omega_x\sin\psi + \Omega_y\cos\psi) - \frac{d\beta}{dt}\frac{d\varphi}{dt}\right] +$$

$$+dM_{ez}^{(inter)} + dM_{ez}^{(exter)} - dmy_b A_x^{(h)}.$$

The left part represents inertia to the element accelerated rotation around the rotor axis. The multiplier $dI_{zz}^{(b)}\cos\beta + dmy_b l_{FH}$ represents the moment of inertia around the rotor axis.

The first summand in the right part is a projection on the normal z-axis at angle $\pi/2 - \beta$ of an inertial force moment due to the accelerated rotation of the rotor system. Despite small values of this angular acceleration, the term is analyzed to figure out the impact of the angular acceleration of the rotor system on the rotor system performance around the z-axis.

The second summand in the right part is a projection on the z-axis of a centrifugal force moment due to the rotor system rotation. This term has an order of magnitude of $\approx dI_{yy}^{(b)}(\Omega_y^2 + \Omega_y^2)$, which is quite small due to small angular velocities of the rotor system rotation ($|\Omega_x| \ll \omega$ and $|\Omega_y| \ll \omega$) and can be neglected.

The third summand in the right part is a projection on the normal z-axis at the angle $\pi/2 - \beta$ of a moment of the Coriolis force due to mutual rotation around the rotor shaft with the flapping motion. The fourth summand in the right part is a moment of the Coriolis force due to mutual rotation of the rotor system with the flapping motion. All other moments due to Coriolis forces, which are collected in square brackets with the multiplier $dI_{Cxx}^{(b)}$, are negligibly small due to the high wing aspect ratio of the blade, which causes $dI_{Cxx}^{(b)} \ll dI_{zz}^{(b)}$, and can be neglected.

The term $dM_{ez}^{(exter)}$ represents a projection on the normal axis of the moments of the external forces (3.28) and has the following components:

$$dM_{ez}^{(exter)} = -y_b dX - dmy_b g_x^{(h)} + M_{z(react)}(y_b).$$

The z-coordinate of the moment of the element aerodynamic force is caused by the tangential component dX of the element aerodynamic force (see 3.5.7) with an arm y_b from the flapping hinge center to the element position. The z-coordinate of the moment of the element weight is represented by the term $-dmy_b g_x^{(h)}$. The element weight is negligibly small in comparison with the aerodynamic forces as assumed in 2.5.1; therefore, the element weight moment can be neglected. The z-component of the moment of the reaction forces is represented only by the flapping hinge reaction moment acting on the root element $M_{z(react)}(0) = M_{FH}$ (see 2.4.11); this component equals zero for all other elements $M_{z(react)}(y_b \neq 0) = 0$.

With all of specified above and with $\sin\beta \approx \beta$ and $\cos\beta \approx 1$, the equation of element rotation around normal z-axis axis is expressed in a simplified form:

$$(dI_{zz}^{(b)} + dmy_b l_{FH})\frac{d\omega}{dt} = -dI_{zz}^{(b)}\beta\left(\frac{d\Omega_x}{dt}\cos\psi - \frac{d\Omega_y}{dt}\sin\psi\right) + 2dI_{Cyy}^{(b)}\omega\beta\frac{d\beta}{dt} -$$

$$- 2dI_{Cyy}^{(b)}\frac{d\beta}{dt}(\Omega_x\cos\psi - \Omega_y\sin\psi) - dM_{ez}^{(inter)} + y_b dX - M_{z(react)}(y_b).$$

$$(3.40)$$

4 Characteristic Measures of Non-uniform Parameters

A blade of the rotor system usually does not have identical aerodynamic properties along its spar; a blade may have twisting; a blade performs flapping motion around its flapping hinge with non-zero offset. Besides these, the induced airflow is inhomogeneous around the rotor disc i.e. has different velocities at different blade azimuthal positions and different blade element positions. All of this makes the analysis of the rotor system properties very complicated and very extended.

A set of measures are introduced in this chapter to simplify, systematize, and generalize the inhomogeneous properties of a blade and induced airflow.

4.1 BASIC CONCEPT

4.1.1   As was stated, chord c_e, lift coefficient slope C_{Le}^{α}, and profile drag coefficient C_{De} of a blade element may be not the same for all elements of the blade and may change with element position along the blade longitudinal y_b-axis in a general case. The dependences of these parameters on element position y_b can be represented by functions $c(y_b)$, $C_L^{\alpha}(y_b)$, and $C_D(y_b)$. Pitch angle $\varphi_e(y_b)$ of a blade element is determined not only by position and orientation of a swashplate but by the blade twisting $\varphi_{tw}(y_b)$, which determines difference of pitch angle of a blade element at position y_b and the blade root element at position $y_b = 0$: $\varphi_e(y_b) = \varphi_e(0) + \varphi_{tw}(y_b)$. Position of an element aerodynamic center may be different for different blade elements; the dependence of transverse position of an element aerodynamic center $x_{ac}^{(b)}$ on the element position y_b can be described by a function $x_{ac}(y_b)$. A pitching moment coefficient of a blade element C_{me} also may change with the element position, and this dependence can be described by a function $C_m(y_b)$. These dependencies $c(y_b)$, $C_L^{\alpha}(y_b)$, $C_D(y_b)$, $\varphi_{tw}(y_b)$, as well as $x_{ac}(y_b)$, $C_m(y_b)$ describe a configuration of a blade in terms of interaction with oncoming airflow and directly depend on geometry and form of the blade: planform of the blade and blade airfoil.

Chord, lift coefficient slope, profile drag coefficient, twisting, pitching moment coefficient, and aerodynamic center position of a blade are called the blade non-uniform parameters hereafter.

4.1.2   Induced airflow, which is caused by rotor vortex sheets created at motion of the rotor blades in the airspace, blows each blade element parallel to the rotor axis. Velocity of a blade element relative to the induced flow is different at different blade element positions y_b and may depend on the blade azimuthal position ψ: $V_i(y_b, \psi)$. In a general case, the induced velocity is inhomogeneous radially along a blade, as well as azimuthally over the rotor disc. Induced velocity $V_i(y_b, \psi)$ blowing an element, which locates in position y_b along its blade with blade azimuthal position ψ, is assumed to be known in the discussed blade element rotor theory.

DOI: 10.1201/9781003296232-4   **63**

In the blade element rotor theory, it is common to use an inflow ratio λ to characterize airflow, which blows a blade element parallelly to the rotor axis and consist of the induced velocity V_i and the rotor system velocity projection on the rotor axis V_y:

$$\lambda(y_b, \psi) := \frac{V_i(y_b, \psi) + V_y}{\omega R}.$$

The λ parameter is different for different blade elements and may differ at different blade azimuthal positions due to the distribution of the induced velocity; therefore, the inflow ratio is represented as a function $\lambda(y_b, \psi)$ on element position along the blade y_b and blade azimuthal position ψ. The induced velocity and the inflow ratio represent the inhomogeneity of the inflow over the rotor disc.

4.1.3   To characterize blade non-uniform properties and the inhomogeneous inflow in a frame of the determination of the rotor system dynamic properties, it is not necessary to know the correspondent functions. As it comes out of a mathematical model of a rotor system based on the discussed blade element rotor theory, which will be further presented, the non-uniform parameters can be characterized by a finite set of measures. The measures are originated as notations to simplify the mathematical derivation; however, it was found that the measures can fully characterize the non-uniform parameters in the frame of the discussed theory. The measures are introduced and systematized in this section prior the further analysis, where they will be logically applied.

4.2   BLADE CHARACTERISTIC PARAMETERS

4.2.1   In order to characterize ability of a blade to create a lift force at rotational motion with non-uniform chord $c(y_b)$ and non-uniform lift coefficient slope $C_L^\alpha(y_b)$, here are introduced blade characteristic parameters $B^{(1)}, B^{(2)}, ...B^{(n)}$. A blade characteristic parameter $B^{(n)}$ of non-zero natural power n ($n \in \mathbf{N} \wedge n \geq 1$) equals:

$$B^{(n)} := \frac{n}{R^{n-1} S_B \bar{C}_L^\alpha} \int_{-l_{FH}}^{L_B} (y_b + l_{FH})^{n-1} c(y_b) C_L^\alpha(y_b) dy_b, \qquad (4.1)$$

$$S_B := \int_{-l_{FH}}^{L_B} c(y_b) dy, \quad \bar{C}_L^\alpha := \frac{1}{R} \int_{-l_{FH}}^{L_B} C_L^\alpha(y_b) dy_b,$$

where S_B is blade planform area, $\bar{C}_L^\alpha$ is an average lift coefficient slope along the blade, and $R = l_{FH} + L_B$ is the radius of the rotor disc.

It must be noted, that the integration is performed not just within the blade $[0, L_B]$ but includes the space in between the blade flapping hinge and the hub center $[-l_{FH}, 0)$. Such integration boundaries allow to evaluate all parts of the rotor disc. It can be assumed that no lift force is created in the area of the hub before the flapping hinge: $\forall y_b \in [-l_{FH}, 0), C_L^\alpha(y_b) = 0$. With this assumption, the blade characteristic parameter can be expressed with integration boundaries, which limit just the blade, $[0, L_B]$:

$$B^{(n)} = \frac{n}{R^{n-1} S_B \bar{C}_L^\alpha} \int_0^{L_B} (y_b + l_{FH})^{n-1} c(y_b) C_L^\alpha(y_b) dy_b.$$

4.2.2   If blade element lift coefficient slope and blade element chord are constant along a blade ($C_L^\alpha(y_b) = const, c(y_b) = const$), then the blade characteristic parameter $B^{(n)}$ equals one with a caveat of a small blade flapping hinge ($l_{FH} \approx 0$):

$$B^{(n)} \Big|_{\substack{C_L^\alpha(y_b)=const \\ c(y_b)=const \\ l_{FH}=0}} = 1.$$

This case describes a rectangular blade with uniform aerodynamic properties and with a negligibly small flapping hinge offset. Such assumptions are applied in the conventional blade element rotor theories.

4.2.3   Most rotor blades usually have similar aerodynamic profiles along their spars but with different element chord lengths. To illustrate the dependence of the blade characteristic parameters on blade planform, an example with a trapezoidal blade and negligibly small flapping hinge offset is presented here. Blade chord changes linearly along a blade spar from $c(0)$ for the root element to $c(L)$ for the tip element. Blade characteristic parameters are expressed for this case as:

$$B^{(n)} = \frac{2}{n+1} \frac{1 + n\frac{c(L)}{c(0)}}{1 + \frac{c(L)}{c(0)}};$$

where $c(L)/c(0)$ is a blade taper ratio.

A triangular planform shape of a blade with $c(L) = 0$ is quantitatively analyzed as a particular case for comparison with a rectangular blade. The characteristic parameters for such a triangular blade are:

$$B^{(n)} = \frac{2}{n+1};$$

or

$$B^{(1)} = 1, \quad B^{(2)} = \frac{2}{3}, \quad B^{(3)} = \frac{1}{2}, \quad B^{(4)} = \frac{2}{5}.$$

There is a distinguished difference in values of the blade characteristic parameters with high powers n for rectangular and triangular blades. The higher is the power n, the larger is the difference. It illustrates sensitivity of these parameters to the blade planform shape.

4.2.4   It was stated the idealistic approach that attack angle of any blade element is less than its stalling angle (see 2.5.6), and the lift force on the blade tip linearly depends on its attack angle. However, the blade lift force vanishes in an area close to the blade tip because of three-dimensional flow of vortex sheets, turbulence flow and unsteady wakes [3]. It causes tip leakage flow when a part of the induced flow near a blade tip returns above the rotor disc and does not participate in rotor thrust generation.

Some approaches of the conventional blade element rotor theory consider such effects that: this blade tip area does not create any lift force; a blade part, which creates lift force, is effectively shorter; the rotor with shorter blades should be used for the calculation of the rotor dynamic properties in order to take into account these effects. A blade tip loss factor B is used for this purpose that shows the part of the rotor disc radius R_{eff}, which effectively participates in the rotor thrust generation, in relation to the whole rotor radius:

$$B := \frac{R_{eff}}{R}.$$

The blade tip loss factor is less than one and usually is in a range of $0.93 - 0.97$.

The blade characteristic parameters can be considered according to the tip loss interpretation; the integration range is shorted in this case:

$$B^{(n)} = \frac{n}{R^{n-1}S_B\bar{C}_L^\alpha} \int_{-l_{FH}}^{BR-l_{FH}} (y_b + l_{FH})^{n-1} c(y_b) C_L^\alpha (y_b) dy_b.$$

Considering a rectangular blade with uniform blade properties ($c(y_b) = const$, $C_L(y_b) = const$), a blade characteristic parameter is determined only by the tip loss factor raised to a power, which equals to the power of the characteristic parameter:

$$B^{(n)} = B^n.$$

It is not correct to associate the blade characteristic parameters only with the tip loss factor. The blade characteristic parameters widely represent the ability of the blade to produce a lift force.

4.2.5   Since the current blade element rotor theory analyzes the influence of flapping hinge offset on rotor system performances, a blade-hinge characteristic parameter $B_m^{(n)}$ with natural m less than n ($m \in \mathbf{N} \wedge 0 \leq m < n$) is introduced:

$$B_m^{(n)} := \frac{n}{R^{n-1}S_B\bar{C}_L^\alpha} \int_{-l_{FH}}^{L_B} y_b^m (y_b + l_{FH})^{n-m-1} c(y_b) C_L^\alpha (y_b) dy_b. \tag{4.2}$$

The parameter $B_m^{(n)}$ quantitatively describes the blade non-uniform properties with the impact of the flapping hinge offset. The parameter $B_m^{(n)}$ is associated with the correspondent blade characteristic parameter $B^{(n)}$. The parameter $B_m^{(n)}$ converges to the correspondent blade characteristic parameter $B^{(n)}$ at the assumption that the flapping hinge offset is much shorter than the blade length:

$$\lim_{l_{FH} \to 0} B_m^{(n)} = B^{(n)}.$$

The parameter $B_m^{(n)}$ can be arithmetically calculated based on blade characteristic parameters $B^{(n)}$. For example:

$$B_1^{(2)} = 2 \left(\frac{B^{(2)}}{2} - \bar{l}_{FH} B^{(1)} \right);$$

$$B_1^{(3)} = 3 \left(\frac{B^{(3)}}{3} - \bar{l}_{FH} \frac{B^{(2)}}{2} \right);$$

$$B_1^{(4)} = 4 \left(\frac{B^{(4)}}{4} - \bar{l}_{FH} \frac{B^{(3)}}{3} \right);$$

$$B_2^{(3)} = 3 \left(\frac{B_1^{(3)}}{3} - \bar{l}_{FH} \frac{B_1^{(2)}}{2} \right) = 3 \left(\frac{B^{(3)}}{3} - \bar{l}_{FH} B^{(2)} + \bar{l}_{FH}^2 B^{(1)} \right);$$

$$B_2^{(4)} = 4 \left(\frac{B_1^{(4)}}{4} - \bar{l}_{FH} \frac{B_1^{(3)}}{3} \right) = 4 \left(\frac{B^{(4)}}{4} - 2\bar{l}_{FH} \frac{B^{(3)}}{3} + \bar{l}_{FH}^2 \frac{B^{(2)}}{2} \right); \qquad (4.3)$$

where $\bar{l}_{FH} := l_{FH}/R$ is normalized flapping hinge offset.

Blade-hinge characteristic parameters simplify the further mathematical expressions.

4.3 TWISTING CHARACTERISTIC PARAMETERS

4.3.1 In order to generalize twisting along a blade $\varphi_{tw}(y_b)$, characteristic twisting angles $\varphi^{(1)}$, $\varphi^{(2)}$, ... $\varphi^{(n)}$ are introduced. A (basic) characteristic twisting angle $\varphi^{(n)}$ of non-zero natural power n represents value which quantitatively generalizes element pitch angle change along a blade:

$$\varphi^{(n)} := \frac{1}{B^{(n)}} \frac{n}{R^{n-1} S_B \bar{C}_L^\alpha} \int_{-l_{FH}}^{L_B} (y_b + l_{FH})^{n-1} \varphi_{tw}(y_b) c(y_b) C_L^\alpha(y_b) dy_b, \qquad (4.4)$$

Similar to the blade characteristic parameter $B^{(n)}$, the integration boundaries include the interval between the hub center and the blade flapping hinge $[-l_{FH}, 0)$, despite that there is no blade and its twisting. This interval is usually excluded by stating a zero lift coefficient slope within this interval.

4.3.2 Without neglecting flapping hinge offset, it will be useful for further analysis to introduce a characteristic twisting angle $\varphi_m^{(n)}$ with natural m less than n ($m \in \mathbf{N} \wedge 0 \leq m < n$), which quantitatively describes the blade twisting with taking into account the blade flapping hinge offset:

$$\varphi_m^{(n)} := \frac{1}{B_m^{(n)}} \frac{n}{R^{n-1} S_B \bar{C}_L^\alpha} \int_{-l_{FH}}^{L_B} y_b^m (y_b + l_{FH})^{n-m-1} \varphi_{tw}(y_b) c(y_b) C_L^\alpha(y_b) dy_b. \qquad (4.5)$$

Such characteristic twisting angle $\varphi_m^{(n)}$ converges to the correspondent $\varphi^{(n)}$ at the assumption of negligibly small flapping hinge offset:

$$\lim_{l_{FH} \to 0} \varphi_m^{(n)} = \varphi^{(n)}.$$

Characteristic twisting angles $\varphi_m^{(n)}$ can be arithmetically expressed in terms of $\varphi^{(n)}$, $B_m^{(n)}$ and $B^{(n)}$; for example:

$$\varphi_1^{(2)} = \frac{2}{B_1^{(2)}} \left(\frac{B^{(2)}}{2} \varphi^{(2)} - \bar{l}_{FH} B^{(1)} \varphi^{(1)} \right);$$

$$\varphi_1^{(3)} = \frac{3}{B_1^{(3)}} \left(\frac{B^{(3)}}{3} \varphi^{(3)} - \bar{l}_{FH} \frac{B^{(2)}}{2} \varphi^{(2)} \right);$$

$$\varphi_1^{(4)} = \frac{4}{B_1^{(4)}} \left(\frac{B^{(4)}}{4} \varphi^{(4)} - \bar{l}_{FH} \frac{B^{(3)}}{3} \varphi^{(3)} \right).$$

4.3.3   The characteristic twisting angles $\varphi^{(n)}$ and $\varphi_m^{(n)}$ will be used further as offsets of collective pitch θ_0 in order to describe the impact of blade twisting as well as to simplify expressions.

4.4   INFLOW RATIO CHARACTERISTIC PARAMETERS

4.4.1   In order to characterize the radially inhomogeneous inflow ratio along a blade at certain azimuthal position of the blade ψ, (basic) characteristic inflow ratios $\lambda^{(1)}(\psi)$, $\lambda^{(2)}(\psi)$, ... $\lambda^{(n)}(\psi)$ are introduced. A characteristic inflow ratio $\lambda^{(n)}(\psi)$ of non-zero natural power n represents value which quantitatively generalizes inflow along a blade in blade azimuthal position ψ and equals:

$$\lambda^{(n)}(\psi) := \frac{n}{R^{n-1} S_B \bar{C}_L^\alpha} \int_{-l_{FH}}^{L_B} (y_b + l_{FH})^{n-1} \lambda(y_b, \psi) c(y_b) C_L^\alpha(y_b) dy_b. \tag{4.6}$$

It is supposed here that airflow exists in the interval between the hub center and the blade flapping hinge: $\exists \lambda(y_b, \psi), y_b \in [-l_{FH}, 0)$. As it was noted in the similar case for $B_m^{(n)}$ and $\varphi_m^{(n)}$ that no lift force is created in this area ($\forall y_b \in [-l_{FH}, 0), C_L^\alpha(y_b) = 0$); therefore, this interval is excluded from calculations.

4.4.2   Without neglecting the flapping hinge offset, a characteristic inflow ratio $\lambda^{(n,m)}(\psi)$ with natural m less than n ($m \in \mathbf{N} \wedge 0 \leq m < n$) is introduced, which quantitatively describe impact of the flapping hinge offset:

$$\lambda^{(n,m)}(\psi) := \frac{n}{R^{n-1} S_B \bar{C}_L^\alpha} \int_{-l_{FH}}^{L_B} y_b^m (y_b + l_{FH})^{n-m-1} \lambda(y_b, \psi) c(y_b) C_L^\alpha(y_b) dy_b. \tag{4.7}$$

The characteristic inflow ratio $\lambda^{(n,m)}(\psi)$ converges to the correspondent $\lambda^{(n)}(\psi)$ with the assumption of negligibly small flapping hinge offset. It is useful to denote hereafter that $\lambda^{(n,0)}(\psi) = \lambda^{(n)}(\psi)$ taking $m = 0$.

4.4.3   A characteristic inflow ratio $\lambda^{(n,m)}(\psi)$ periodically depends on blade azimuthal position ψ and repeats every blade turn around the rotor shaft. Therefore, the dependence of the characteristic inflow ratio $\lambda^{(n,m)}(\psi)$ on azimuthal position can be represented by Fourier series:

$$\lambda^{(n,m)}(\psi) = \lambda_0^{(n,m)} + \sum_{k=1}^{\infty} \left(\lambda_{ak}^{(n,m)} \cos k\psi + \lambda_{bk}^{(n,m)} \sin k\psi \right),$$

where $\lambda_0^{(n,m)}$ represents an average characteristic inflow ratio of power n and m over the rotor disc:

$$\lambda_0^{(n,m)} = \frac{1}{2\pi} \int_0^{2\pi} \lambda^{(n,m)}(\psi) d\psi;$$

and Fourier coefficients $\lambda_{ak}^{(n,m)}$ and $\lambda_{bk}^{(n,m)}$ represent a k-th harmonic of azimuthal inhomogeneity of the characteristic inflow ratio $\lambda^{(n,m)}(\psi)$ over the rotor disc:

$$\lambda_{ak}^{(n,m)} = \frac{1}{\pi} \int_0^{2\pi} \lambda^{(n,m)}(\psi) \cos(k\psi) d\psi, \quad \lambda_{bk}^{(n,m)} = \frac{1}{\pi} \int_0^{2\pi} \lambda^{(n,m)}(\psi) \sin(k\psi) d\psi.$$

$$(4.8)$$

The set of Fourier coefficients of characteristic inflow ratios $\lambda_0^{(n,m)}$, $\lambda_{ak}^{(n,m)}$, $\lambda_{bk}^{(n,m)}$ with harmonics $k = 1, 2...$, with powers $n = 1, 2...$ and with $0 \leq m < n$ characterizes the inhomogeneous inflow over the rotor disc in the frame of the discussed blade element rotor theory. The index a denotes the Fourier coefficients of harmonics with cosines, which describe longitudinal inflow inhomogeneity over the disc; the index b denotes the Fourier coefficients of harmonics with sines, which describe lateral (sidewise) inflow inhomogeneity.

As a notation simplification, the Fourier coefficients of the characteristic inflow ratio with $m = 0$ are denoted without m:

$$\lambda_0^{(n)} := \lambda_0^{(n,0)}, \ \lambda_{ak}^{(n)} := \lambda_{ak}^{(n,0)}, \ \lambda_{bk}^{(n)} := \lambda_{bk}^{(n,0)}.$$

4.4.4   Some calculations of the rotor dynamic properties use generalization of the inhomogeneity of the inflow ratio around the rotor disc associated with the blade twisting, which is characterized by $\lambda_0^{tw(n)}$, $\lambda_{ak}^{tw(n)}$, and $\lambda_{bk}^{tw(n)}$. These characteristics represent Fourier coefficients of inflow ratio inhomogeneity associated with the twisting:

$$\lambda_0^{tw(n)} := \frac{n}{2\pi} \int\limits_0^{2\pi} \int\limits_{-l_{FH}}^{L_B} \lambda(y_b, \psi) \varphi_{tw}(y_b) \frac{c(y_b) C_L^\alpha(y_b)}{S_B \bar{C}_L^\alpha} \frac{(y_b + l_{FH})^{n-1}}{R^{n-1}} dy d\psi,$$

$$\lambda_{ak}^{tw(n)} := \frac{n}{\pi} \int\limits_0^{2\pi} \cos(k\psi) \int\limits_{-l_{FH}}^{L_B} \lambda(y_b, \psi) \varphi_{tw}(y_b) \frac{c(y_b) C_L^\alpha(y_b)}{S_B \bar{C}_L^\alpha} \frac{(y_b + l_{FH})^{n-1}}{R^{n-1}} dy_b d\psi,$$

$$\lambda_{bk}^{tw(n)} := \frac{n}{\pi} \int\limits_0^{2\pi} \sin(k\psi) \int\limits_{-l_{FH}}^{L_B} \lambda(y_b, \psi) \varphi_{tw}(y_b) \frac{c(y_b) C_L^\alpha(y_b)}{S_B \bar{C}_L^\alpha} \frac{(y_b + l_{FH})^{n-1}}{R^{n-1}} dy_b d\psi,$$

where n is a power these characteristics, $\lambda_0^{tw(n)}$ is a Fourier constant, $\lambda_{ak}^{tw(n)}$ and $\lambda_{bk}^{tw(n)}$ are cosine and sinus Fourier coefficients of k-th harmonics. These coefficients are generally used for the analysis of induced drag acting on a blade.

4.4.5   Fourier coefficients of square of the inflow ratio are seldom used:

$$\Lambda_0^2 := \frac{1}{\pi} \int\limits_0^{2\pi} \int\limits_{-l_{FH}}^{L_B} \frac{c(y_b) C_L^\alpha(y_b)}{S_B \bar{C}_L^\alpha} (\lambda(y_b, \psi))^2 \frac{y_b + l_{FH}}{R} dy_b d\psi,$$

$$\Lambda_a^2 := \frac{1}{\pi} \int\limits_0^{2\pi} \cos\psi \int\limits_{-l_{FH}}^{L_B} \frac{c(y_b) C_L^\alpha(y_b)}{S_B \bar{C}_L^\alpha} (\lambda(y_b, \psi))^2 dy_b d\psi,$$

$$\Lambda_b^2 := \frac{1}{\pi} \int\limits_0^{2\pi} \sin\psi \int\limits_{-l_{FH}}^{L_B} \frac{c(y_b) C_L^\alpha(y_b)}{S_B \bar{C}_L^\alpha} (\lambda(y_b, \psi))^2 dy_b d\psi.$$

4.4.6   It is assumed hereafter, that the second and higher harmonics of azimuthal inhomogeneity of the inflow are negligibly small in comparison with magnitude of the first harmonics; therefore, the Fourier coefficients of second and higher harmonics of these inflow ratio characteristics are neglected hereafter:
$$\lambda_{ak}^{(n,m)} \approx 0, \quad \lambda_{bk}^{(n,m)} \approx 0, \quad \lambda_{bk}^{tw(n)} \approx 0, \quad \lambda_{bk}^{tw(n)} \approx 0 \quad \text{at } k > 2.$$

4.5   CHARACTERISTIC BLADE DRAG COEFFICIENT

In order to characterize the blade profile drag at blade rotational motion with the non-uniform profile drag coefficient $C_D(y_b)$ along the blade, characteristic (profile) drag coefficients $\bar{C}_D^{(1)}, \bar{C}_D^{(2)}, \dots \bar{C}_D^{(n)}$ are introduced. A characteristic (blade profile) drag coefficient $C_D^{(n)}$ of non-zero natural power n equals:

$$\bar{C}_D^{(n)} := \frac{n}{R^{n-1} S_B} \int\limits_{-l_{FH}}^{L_B} (y_b + l_{FH})^{n-1} c(y_b) C_D(y_b) dy_b. \tag{4.9}$$

These parameters represent ability of a blade to produce friction forces with air at blade rotational motion. These parameters are used in the analysis of total rotor forces and moments acting on the rotorcraft.

4.6 MEASURES ASSOCIATED WITH AERODYNAMIC CENTER POSITION

Transverse position of an element aerodynamic center basically represents an arm of element aerodynamic forces creating a moment around the longitudinal axis. A sum of the moments along a blade is compensated by a moment created by a blade pitch force on the blade pitch horn. Thus, the total aerodynamic moment of the blade around the longitudinal axis basically determines the blade pitch force. Characteristic parameters associated with aerodynamic center change along the blade are introduced for a purpose of blade pitch force determination despite perfunctory analyzing of the blade pitch force hereafter.

4.6.1   In order to generalize transverse position of element aerodynamic center $x_{ac}(y_b)$ along the blade, characteristic normalized positions of blade aerodynamic center $\bar{x}_{ac}^{(1)}$, $\bar{x}_{ac}^{(2)}$, ... $\bar{x}_{ac}^{(n)}$ are introduced. A characteristic normalized position of aerodynamic center of a blade $\bar{x}_{ac}^{(n)}$ of non-zero natural power n represents value which quantitatively describes variation of transverse aerodynamic center position along the blade, which is normalized on the rotor disc radius (R):

$$\bar{x}_{ac}^{(n)} := \frac{1}{B^{(n)}} \frac{n}{R^n} \int_{-l_{FH}}^{L_B} (y_b + l_{FH})^{n-1} \frac{x_{ac}(y_b)c(y_b)C_L^{\alpha}(y_b)}{S_B \bar{C}_L^{\alpha}} dy_b.$$

4.6.2   The characteristic twisting angle $\varphi_{xad}^{(n)}$ of non-zero natural power n represents value which quantitatively generalizes impact of the blade twisting associated with the blade aerodynamic center change on the blade pitch force:

$$\varphi_{xad}^{(n)} := \frac{1}{B^{(n)}\bar{x}_{ac}^{(n)}} \frac{n}{R^n} \int_{-l_{FH}}^{L_B} (y_b + l_{FH})^{n-1} \frac{x_{ac}(y_b)\varphi_{tw}(y_b)c(y_b)C_L^{\alpha}(y_b)}{S_B \bar{C}_L^{\alpha}} dy_b,$$

4.6.3   A characteristic parameter $\lambda^{ac(n)}(\psi)$ of non-zero natural power n describes mutual impact of the non-uniform inflow and the blade aerodynamic center change along the blade on the blade pitch force in blade azimuthal position ψ and equals:

$$\lambda^{ac(n)}(\psi) := \frac{n}{R^n} \int_{-l_{FH}}^{L_B} (y_b + l_{FH})^{n-1} \frac{x_{ac}(y_b)\lambda(y_b, \psi)c(y_b)C_L^{\alpha}(y_b)}{S_B \bar{C}_L^{\alpha}} dy_b.$$

4.6.4   A characteristic parameter $\lambda^{ac(n)}(\psi)$ periodically depends on the blade azimuthal position and can be decomposed into a Fourier series:

$$\lambda^{ac(n)}(\psi) = \lambda_0^{ac(n)} + \sum_{k=1}^{\infty} \left(\lambda_{ak}^{ac(n)} \cos k\psi + \lambda_{bk}^{ac(n)} \sin k\psi \right),$$

$$\lambda_0^{ac(n)} := \frac{1}{2\pi} \int_0^{2\pi} \lambda^{ac(n)}(\psi)\,d\psi,$$

$$\lambda_{ak}^{ac(n)} := \frac{1}{\pi} \int_0^{2\pi} \lambda^{ac(n)}(\psi)\cos(k\psi)\,d\psi, \quad \lambda_{bk}^{ac(n)} := \frac{n}{\pi} \int_0^{2\pi} \lambda^{ac(n)}(\psi)\sin(k\psi)\,d\psi,$$

where $\lambda_0^{ac(n)}$ is average parameter of $\lambda^{ac(n)}(\psi)$ over the rotor disc; $\lambda_{ak}^{ac(n)}$, and $\lambda_{ak}^{ac(n)}$ are cosine and sine Fourier coefficients of a k-th harmonic of the $\lambda^{ac(n)}(\psi)$.

4.7 CHARACTERISTIC PITCHING MOMENT COEFFICIENT

In order to generalize the pitching moment coefficient $C_m(y_b)$ along a blade, characteristic pitching moments $m_{ad}^{(1)}, m_{ad}^{(2)}, \dots m_{ad}^{(n)}$ are introduced. A characteristic pitching moment of a blade $m_{ad}^{(n)}$ of non-zero natural power n equals:

$$m_{ac}^{(n)} := \frac{1}{B^{(n)}} \frac{n}{R^n S_B \bar{C}_L^\alpha} \int_{-l_{FH}}^{L_B} (y_b + l_{FH})^{n-1} C_m(y_b)(c(y_b))^2\,dy_b.$$

Similarly to characteristic positions of aerodynamic center, these parameters are generally used for determination of a blade pitch force. These parameters usually are used as offsets of collective pitch in expressions of blade pitch forces.

5 Analysis of Blade Rotations

Rotor blades create lift forces during rotation around the rotor shaft, which participate in the turning of the blades around their flapping hinges. The blade flapping motion depends on the flight condition and the swashplate position and tilt. The blade aerodynamic forces, which are transferred to the rotorcraft, depend on blade rotation around the rotor shaft accompanied by the blade flapping motion. The blade flapping motion is one of the key features to determine the dynamic properties of the rotor system.

The flapping motion of a blade during rotation around the rotor shaft as well as associated properties with these rotations are analyzed in this chapter.

5.1 BLADE ROTATION EQUATIONS

5.1.1 Rotation equations of all elements of a blade are generalized here assuming, that the blade is an absolutely rigid body. All elements, which belong to a single rigid blade, perform same rotations and have same angular velocities. This enables to sum all of the rotation equations (3.34) of elements, which belong to the same blade. A resultant equation in a vector form describes rotations of the whole blade:

$$
\begin{aligned}
I\left(-\frac{d^2\beta}{dt^2}\vec{e}_x + \frac{d^2\varphi}{dr^2}\vec{e}_y + \vec{\varepsilon}_\omega^{(h)} + \vec{\varepsilon}_\Omega^{(h)}\right) + J\vec{e}_y \times \left((\vec{\varepsilon}_\omega^{(h)} + \vec{\varepsilon}_\Omega^{(h)}) \times \vec{l}_{FH}^{(h)}\right) = \\
= \sum_{m=1}^{Ne} d\vec{M}_{em}^{(inter)} + \sum_{m=1}^{Ne} d\vec{M}_{em}^{(exter)} - J\vec{e}_y \times \vec{A}^{(h)} - \\
- \left[\vec{\omega}^{(h)} \times I\vec{\omega}^{(h)} + \vec{\Omega}^{(h)} \times I\vec{\Omega}^{(h)} + \left(\frac{d\beta}{dt}\right)^2 \vec{e}_x \times I\vec{e}_x + \left(\frac{d\varphi}{dt}\right)^2 \vec{e}_y \times I\vec{e}_y + \right. \\
\left. + J\vec{e}_y \times \left(\vec{\omega}^{(h)} \times (\vec{\omega}^{(h)} \times \vec{l}_{FH}^{(h)})\right) + J\vec{e}_y \times \left(\vec{\Omega}^{(h)} \times (\vec{\Omega}^{(h)} \times \vec{l}_{FH}^{(h)})\right) \right] + \\
+ \left\{2\left(-\frac{d\beta}{dt}\vec{e}_x + \frac{d\varphi}{dr}\vec{e}_y\right) \times I_C(\vec{\omega}^{(h)} + \vec{\Omega}^{(h)}) + 2\vec{\omega}^{(h)} \times I_C\vec{\Omega}^{(h)} - \right. \\
\left. - 2\frac{d\beta}{dt}\frac{d\varphi}{dt}\vec{e}_y \times I_C\vec{e}_x - 2J\vec{e}_y \times \left(\vec{\Omega}^{(h)} \times \left(\vec{\omega}^{(h)} \times \vec{l}_{FH}\right)\right)\right\}.
\end{aligned}
\tag{5.1}
$$

The term J represents a mass moment of the blade around its flapping hinge:

$$
J = \int_0^{L_B} y_b dm.
$$

The mass moment is determined by position of the blade center of mass y_{Bmc} on the blade longitudinal y_b-axis and by the total blade mass m_B: $J = m_B y_{Bmc}$.

DOI: 10.1201/9781003296232-5

The total moment of element inertial forces due to accelerated motion of the whole rotor system $(-J\vec{e}_y \times \vec{A}^{(h)})$ can be neglected based on the assumption of insignificant acceleration of the rotorcraft (see 2.5.2).

5.1.2   The tensor I represents moments of inertia of the blade in the blade-hinge frame in a tensor form. Based on the assumption of small value of blade pitch angle similarly, as for the blade element inertia tensor in 3.8.6, the blade inertia tensor in the blade-hinge frame approximately equals to the blade inertia tensor in the blade fixed frame $I^{(b)}$:

$$I \approx I^{(b)} = \begin{bmatrix} I_{xx}^{(b)} & 0 & 0 \\ 0 & I_{yy}^{(b)} & 0 \\ 0 & 0 & I_{zz}^{(b)} \end{bmatrix}, \tag{5.2}$$

where $I_{xx}^{(b)}$, $I_{yy}^{(b)}$, $I_{zz}^{(b)}$ are blade moments of inertia around the correspondent axis of the blade fixed frame and remain constant. A blade moment of inertia around each axis is the sum of the moments of inertia of all blade elements around this axis (3.35):

$$I_{xx}^{(b)} = \int_0^{L_B} dI_{xx}^{(b)} = \sum_{j=0}^{N_p} m_j(y_{bj}^2 + z_{bj}^2),$$

$$I_{yy}^{(b)} = \int_0^{L_B} dI_{yy}^{(b)} = \sum_{j=0}^{N_p} m_j(x_{bj}^2 + z_{bj}^2),$$

$$I_{zz}^{(b)} = \int_0^{L_B} dI_{zz}^{(b)} = \sum_{j=0}^{N_p} m_j(x_{bj}^2 + y_{bj}^2),$$

where N_p is the total number of all point particles in the blade, j is an index of a point particle with mass m_j located in $\vec{p}_j^{(b)}(x_{bj}, y_{bj}, z_{bj})$ with coordinates the blade fixed frame.

Here must be noted as a remark that the blade moments of inertia around the blade transverse x_b-axis ($I_{xx}^{(b)}$) and around the blade normal z_b-axis ($I_{zz}^{(b)}$) are not the blade central principal moments of inertia, which are considered around blade central principal axes passing the blade center of mass. The blade moment of inertia $I_{yy}^{(b)}$ around the blade longitudinal y_b-axis can be considered as a central principal moment of inertia of the blade; however, it is not the point of this context.

With this diagonal tensor of the blade moment of inertia, the moments of the centrifugal forces acting on the blade due to the flapping motion and the moments of the centrifugal forces due to rotation around the feathering hinge equal zero:

$$\left(\frac{d\beta}{dt}\right)^2 \vec{e}_x \times I\vec{e}_x = 0, \quad \left(\frac{d\varphi}{dt}\right)^2 \vec{e}_y \times I\vec{e}_y = 0.$$

The tensor for blade Coriolis moments I_C is the sum of diagonal tensors for the element Coriolis moments (3.36) of all blade elements:

$$I_C = \begin{bmatrix} I_{Cxx}^{(b)} & 0 & 0 \\ 0 & I_{Cyy}^{(b)} & 0 \\ 0 & 0 & I_{Czz}^{(b)} \end{bmatrix}, \tag{5.3}$$

where

$$I_{Cxx}^{(b)} = \int_0^{L_B} dI_{Cxx}^{(b)} = \sum_{j=0}^{Np} m_j x_{bj}^2,$$

$$I_{Cyy}^{(b)} = \int_0^{L_B} dI_{Cyy}^{(b)} = \sum_{j=0}^{Np} m_j y_{bj}^2,$$

$$I_{Czz}^{(b)} = \int_0^{L_B} dI_{Czz}^{(b)} = \sum_{j=0}^{Np} m_j z_{bj}^2.$$

Essentially, the tensor of blade moment of inertia is the sum of tensors of inertia moment of all blade elements; the tensors for Coriolis moments is the sum of tensors for Coriolis moment of all blade elements.

5.1.3   The sum of all internal torques of all blade elements is represented in (5.1) by the sum $\sum_{m=1}^{Ne} d\vec{M}_{em}^{(inter)}$, where N_e is the number of all elements in the blade. This sum represents interactions between point particles that do not belong to same element:

$$\sum_{m=1}^{Ne} d\vec{M}_{em}^{(inter)} = \sum_{m=1}^{Ne} \sum_{i \in m} \sum_{j \notin m} \vec{p}_i \times \vec{F}_{i,j}^{(inter)},$$

where i is a point particle which belongs to a certain m-th blade element, and j is a point particle which belongs to any other blade element except the m-th blade element. This nested sum can be reordered with respect to a couple of the interacting particles, each of which belongs to a different element:

$$\sum_{m=1}^{Ne} d\vec{M}_{em}^{(inter)} = \sum_{i=0}^{Np} \sum_{\substack{j=0, \\ j>i \wedge i \in m \wedge j \notin m}}^{Np} \left(\vec{p}_i \times \vec{F}_{i,j}^{(inter)} + \vec{p}_j \times \vec{F}_{j,i}^{(inter)} \right) = 0.$$

According to the Newton's third law, forces ($\vec{F}_{i,j}$ and $\vec{F}_{j,i}$) of an interacting particle couple have same magnitude, act along a line, which joins these particles, but in opposite directions. A sum of force moments of such couple equals zero as moments of two equal and opposite forces acting along one line: $\vec{p}_i \times \vec{F}_{i,j}^{(inter)} + \vec{p}_j \times \vec{F}_{j,i}^{(inter)} = 0$. Therefore, the whole nested sum, i.e. a sum of internal torques of all blade elements, equals zero.

5.1.4   The sum of moments of external element forces over the whole blade, which is represented by the second summand in the right part of (5.1), includes all external force moments acting on each element in the blade and can be rewritten as a vector sum of all external moments based on the definition of $\vec{M}_e^{(exter)}$ in (3.28):

$$\sum_{m=1}^{Ne} d\vec{M}_{em}^{(exter)} = J\vec{e}_y \times \vec{g}^{(h)} + \sum_{m=1}^{Ne} \left(\vec{p}_{acm} \times d\vec{R}_m + \vec{e}_y dM_{ADm} \right) + \vec{e}_z M_{FH} + \vec{M}_{pitch},$$

where: $J\vec{e}_y \times \vec{g}^{(h)}$ is a moment of the blade weight about flapping hinge center; the second term with the sum is total moment of blade aerodynamic forces; $\vec{e}_z M_{FH}$ is the flapping hinge reaction moment defined in 2.4.11; $\vec{M}_{pitch}$ is the moment of the blade pitch force defined in 3.8.3. The total weight of blades is much smaller than the total aerodynamic force created by the blades as assumed in 2.5.1; therefore, the moment of the blade weight term $J\vec{e}_y \times \vec{g}^{(h)}$ is neglected here.

5.1.5   It is useful to present the blade rotation equation (5.1) as a system of equations, which consists of equations for each coordinate of the blade-hinge frame. Each blade rotation equation for a certain coordinate can be built as a sum of rotation equations of all blade elements for the correspondent coordinate found in (3.37), (3.39), and (3.40). The system of blade rotation equations for the coordinates can be represented in the following way with the zero sum of the internal torques, in terms of the defined blade moments of inertia $I_{xx}^{(b)}$, $I_{yy}^{(b)}$, $I_{zz}^{(b)}$, with the defined elements of the tensor for blade Coriolis moments $I_{Cxx}^{(b)}$, $I_{Cyy}^{(b)}$, $I_{Czz}^{(b)}$, and with the blade mass moment J:

$$
\left\{
\begin{aligned}
&I_{xx}^{(b)}\frac{d^2\beta}{dt^2} = (I_{xx}^{(b)} + Jl_{FH})\left(\frac{d\Omega_x}{dt}\sin\psi + \frac{d\Omega_y}{dt}\cos\psi\right) - \\
&- (I_{zz}^{(b)} + Jl_{FH} - I_{yy}^{(b)})\omega^2\beta + \\
&+ 2\left(I_{Cyy}^{(b)} + Jl_{FH}\right)\omega(\Omega_x\cos\psi - \Omega_y\sin\psi) - \\
&- 2kI_{Czz}^{(b)}\frac{d\beta}{dt}\omega + 2I_{Czz}^{(b)}\omega^2(\theta_1\sin\psi - \theta_2\cos\psi) - \sum_{m=1}^{Ne} y_b dZ - y_{pitch}^{(b)}R_{pitch} \\[2mm]
&I_{yy}^{(b)}k\frac{d^2\beta}{dt^2} = -dI_{yy}^{(b)}\left(\frac{d\Omega_x}{dt}\cos\psi - \frac{d\Omega_y}{dt}\sin\psi\right) \\
&+ I_{yy}^{(b)}\omega^2(\theta_1\cos\psi + \theta_2\sin\psi) + 2I_{Cxx}^{(b)}\omega(\Omega_x\sin\psi + \Omega_y\cos\psi) + \\
&+ 2I_{Czz}^{(b)}\frac{d\beta}{dt}\omega + \sum_{m=1}^{Ne} x_{ac}dZ - \sum_{m=1}^{Ne} dM_{AD} + x_{pitch}^{(b)}R_{pitch} \\[2mm]
&(I_{zz}^{(b)} + Jl_{FH})\frac{d\omega}{dt} = -dI_{zz}^{(b)}\beta\left(\frac{d\Omega_x}{dt}\cos\psi - \frac{d\Omega_y}{dt}\sin\psi\right) + 2I_{Cyy}^{(b)}\omega\beta\frac{d\beta}{dt} - \\
&- 2I_{Cyy}^{(b)}\frac{d\beta}{dt}(\Omega_x\cos\psi - \Omega_y\sin\psi) + \sum_{m=1}^{Ne} y_b dX - M_{FH},
\end{aligned}
\right.
$$

$$(5.4)$$

Assuming that every blade element has infinitely small width ($dy_b \to 0$) that causes infinitely large number of elements in the blade ($Ne \to \infty$), then the sum of correspondent coordinates of moments of the element aerodynamic forces applied to the element aerodynamic centers can be represented by integration:

$$
\lim_{\substack{Ne\to\infty \\ dy\to 0}}\sum_{m=1}^{Ne} y_b dY = \int_0^{L_B} y_b dZ, \quad \lim_{\substack{Ne\to\infty \\ dy\to 0}}\sum_{m=1}^{Ne} x_{ad}dZ = \int_0^{L_B} x_{ad}dZ, \quad \lim_{\substack{Ne\to\infty \\ dy\to 0}}\sum_{m=1}^{Ne} y_b dX = \int_0^{L_B} y_b dX.
$$

The total aerodynamic pitch moment for the y-coordinate equation can be represented as well by integration:

$$
\lim_{\substack{Ne\to\infty \\ dy\to 0}}\sum_{m=1}^{Ne} dM_{AD} = \int_0^{L_B} dM_{AD}.
$$

5.1.6   The equation system (5.4) has three unknown parameters: the blade flapping angle β, the blade pitch force R_{pitch}, and the flapping hinge reaction moment M_{FH}. To find these unknown parameters is the task of this section.

5.2   BLADE FLAPPING MOTION

5.2.1   Blade flapping angle dependence on time $\beta(t)$ describes the motion of a blade around its flapping hinge. The flapping angle, which depends on time, is one of the unknown parameters of the system of the blade rotation equations (5.4). In order to get an equation with respect to only this unknown parameter, the first equation of the system (5.4) is added to the second equation, which is previously multiplied on the flapping compensation coefficient k. The coefficient k is determined as the ratio of the longitudinal position of the pitch horn joint with the pitch link ($y_{pitch}^{(b)}$) to the transverse position of this joint ($x_{pitch}^{(b)}$): $k = y_{pitch}^{(b)}/x_{pitch}^{(b)}$. The term of the blade pitch force moment $-y_{pitch}^{(b)}R_{pitch}$ and the term $2kI_{Czz}^{(b)}\omega d\beta/dt$ are reduced after this sum of the equations. The result of this manipulation is a second order differential equation of the blade flapping angle with respect to time, which describes the blade flapping motion:

$$(I_{xx}^{(b)} + I_{yy}^{(b)}k^2)\frac{d^2\beta}{dt^2} =$$

$$= \left[(I_{xx}^{(b)} + Jl_{FH})\left(\frac{d\Omega_x}{dt}\sin\psi + \frac{d\Omega_y}{dt}\cos\psi\right) - kdI_{yy}^{(b)}\left(\frac{d\Omega_x}{dt}\cos\psi - \frac{d\Omega_y}{dt}\sin\psi\right)\right] -$$

$$- (I_{zz}^{(b)} + Jl_{FH} - I_{yy}^{(b)})\beta\omega^2 + 2\left(I_{Cyy}^{(b)} + Jl_{FH}\right)\omega(\Omega_x\cos\psi - \Omega_y\sin\psi) +$$

$$+ 2kI_{Cxx}^{(b)}\omega(\Omega_x\sin\psi + \Omega_y\cos\psi) - \int_0^{L_B}(y_b - kx_{ac})dZ - \int_0^{L_B}kdM_{AD} +$$

$$+ kI_{yy}^{(b)}\omega^2(\theta_1\cos\psi + \theta_2\sin\psi) + 2I_{Czz}^{(b)}\omega^2(\theta_1\sin\psi - \theta_2\cos\psi). \tag{5.5}$$

The blade moment of inertia around the tangential axis $I_{xx}^{(b)}$ is much greater than the blade moment of inertia around the longitudinal axis $I_{yy}^{(b)}$ due to the high wing aspect ratio of the blade. Therefore, the multiplier expression in the parentheses in the left part of the equation can be approximated: $I_{xx}^{(b)} + I_{yy}^{(b)}k^2 \approx I_{xx}^{(b)}$.

The first summand in the right part (the expression within the square brackets) of the equation (5.5) represents moments of inertial forces due to accelerated rotation of the whole rotor system. The first term in the square brackets represents the inertial forces due to accelerated rotation of the rotor system creating a moment around an axis, which is parallel to the tangential x-axis of the blade-hinge frame and passes the rotor hub center. The term $I_{xx}^{(b)} + Jl_{FH}$ represents a moment of the blade inertia around this axis and is denoted $I_{xx}^{(r)}$:

$$I_{xx}^{(r)} := I_{xx}^{(b)} + Jl_{FH}. \tag{5.6}$$

This moment of blade inertia is related to the inertial forces due to rotor system rotation ($\vec{\Omega}$), which is considered about the hub center. In a case of negligibly small offset of blade flapping hinge, the term $I_{xx}^{(r)}$ approximately equals the moment of blade inertia $I_{xx}^{(b)}$ around the tangential x-axis. The second term in the square brackets represents inertial forces due to accelerated rotation of the rotor system creating a moment around the longitudinal y-axis; this term is delivered from the equation around the longitudinal y-axis as a result of the manipulation. This second term is much smaller in comparison with the first term due to the high wing aspect ratio of the blade, that causes $I_{yy}^{(b)} \ll I_{xx}^{(b)}$; therefore, the second term is neglected. Despite the rotor system angular acceleration being quite small, the first term is not neglected and is analyzed hereafter in order to evaluate the impact of this angular acceleration.

The second summand in the right part of the equation (5.5) represents a total moment of centrifugal forces of all blade elements due to rotation around the rotor axis. These centrifugal forces are perpendicularly directed from the rotor axis; therefore, the moment depends on blade flapping angle. If the blade lays on the rotor disc, then the centrifugal forces do not create any moment; if the blade has non-zero flapping angle, then the centrifugal forces create a moment proportionality to the sinus of the flapping angle ($\sin \beta \approx \beta$) and are directed to reduce the flapping. The term $I_{zz}^{(b)} + Jl_{FH}$ represents a blade moment of inertia around the rotor axis (the z_r-axis) and is denoted $I_{zz}^{(r)}$:

$$I_{zz}^{(r)} := I_{zz}^{(b)} + Jl_{FH}. \tag{5.7}$$

This moment of inertia around the rotor axis is related to the centrifugal forces due to rotation around the rotor axis.

The term $I_{zz}^{(b)} + Jl_{FH} - I_{yy}^{(b)}$ represents difference between the moment of blade inertia around the rotor axis $I_{zz}^{(r)}$ and the moment of blade inertia around the blade longitudinal y-axis $I_{yy}^{(b)}$. The term $I_{yy}^{(b)}$ is much smaller than the moment of blade inertia around the normal z-axis due to the high wing aspect ratio of the blade: $I_{zz}^{(b)} \ll I_{yy}^{(b)}$. So that, this term can be approximated $I_{zz}^{(b)} + Jl_{FH} - I_{yy}^{(b)} \approx I_{zz}^{(r)}$.

The third summand in the right part of the equation (5.5) represents a moment of Coriolis forces due to the mutual rotation of the rotor system with the blade rotation around the rotor shaft. The element $I_{Cyy}^{(b)}$ of the tensor for Coriolis moments has value, which is close to the moment of blade inertia around the tangential x-axis $I_{xx}^{(b)}$ due to the high wing aspect ratio of the blade:

$$I_{xx}^{(b)} = \sum_{j=0}^{Np} m_j(y_{bj}^2 + z_{bj}^2) \approx \sum_{j=0}^{Np} m_j y_{bj}^2 = I_{Cyy},$$

where $y_{bj} \gg z_{bj}$ for majority of blade point particles. With this, the multiplier $I_{Cyy}^{(b)} + Jl_{FH}$ of the third summand is close to the moment of blade inertia around the axis which is parallel to the tangential x-axis and passes the hub center: $I_{Cyy}^{(b)} + Jl_{FH} \approx I_{xx}^{(r)}$.

The fourth summand represents a moment around the longitudinal axis created by Coriolis forces due to mutual rotor system rotation with rotation around the shaft, which was delivered from the rotation equation for the y-axis after the equation manipulation. The element $I_{Cxx}^{(b)}$ of the tensor for Coriolis moments has much smaller value than $I_{Cyy}^{(b)}$ due to the high wing aspect ratio of the blade. Therefore, the fourth summand in the right part of the equation with the multiplier $I_{Cxx}^{(b)}$ is much smaller than the third summand in the right part and can be neglected.

Position of an aerodynamic center of a blade element is usually located quite close to the blade longitudinal axis. The transverse coordinate of an element aerodynamic center x_{ac} is much smaller than the longitudinal coordinate of the element y_b for majority of blade elements: $y_b \ll x_{ac}$. Therefore, the term $y_b - kx_{ac}$ can be approximated to y_b in the fifth summand in the equation right part, which mostly represents the total moment of aerodynamic forces of all blade elements around the blade flapping hinge.

An element pitch moment dM_{AD} has an order of magnitude, which does not exceed the transverse moment of the element lift force with an arm equal to the element chord $|dM_{AD}| < |c_e dZ|$: this assumption is based on the fact that an element center of aerodynamic pressure is located within the element volume. Therefore, the element aerodynamic pitch moment dM_{AD} is much smaller than the moment of the element lift force around the flapping hinge for majority of blade elements due to the high wing aspect ratio of the blade: $|dM_{AD}| < |c_e dZ| \ll |y_b dZ|$ at $c_e \ll y_b$. So that, the total pitch moment of all blade elements, which is represented by the sixth summand in the right part (5.5), is negligibly small in comparison to the fifth summand and can be neglected.

The two last summands in the left part of the equation (5.5) represent inertia of the rotation around the feathering hinge, which is transferred via the blade pitch force moment, and a moment of Coriolis forces due to rotation of the blade around the feathering hinge accompanied with the rotation around the rotor shaft. These summands are quite small in comparison to other summands because the moment of inertia $I_{yy}^{(b)}$ and the element of Coriolis tensor $I_{Czz}^{(b)}$ are quite small for a blade with a high wing aspect ratio, and these summands can be neglected.

The blade flapping equation (5.5) is expressed in a simplified form with all of the specified above:

$$
I_{xx}^{(b)} \frac{d^2\beta}{dt^2} = I_{xx}^{(r)} \left(\frac{d\Omega_x}{dt} \sin\psi + \frac{d\Omega_y}{dt} \cos\psi \right) - I_{zz}^{(r)} \beta \omega^2 +
$$
$$
+ 2I_{xx}^{(r)} \omega (\Omega_x \cos\psi - \Omega_y \sin\psi) - \int_0^{L_B} y_b dZ. \tag{5.8}
$$

5.2.2 The blade moment of inertia around the rotor axis $I_{zz}^{(r)}$ and the moment of inertia $I_{xx}^{(r)}$ around the axis, which passes the hub center and is parallel to the tangential x-axis, are quite close due to the high wing aspect ratio of the blade and small flapping angle. Therefore, these two terms are considered approximately equal hereafter: $I_{xx}^{(r)} \approx I_{zz}^{(r)}$.

Here is introduced a ratio $\epsilon := I_{xx}^{(b)}/I_{zz}^{(r)}$, which is analyzed in terms of the flapping hinge offset l_{FH}, the blade mass moment J, and the blade moment of inertia around the blade normal z_b-axis $I_{zz}^{(b)}$ associated with $I_{zz}^{(r)}$ according to (5.7):

$$\epsilon = \frac{I_{xx}^{(b)}}{I_{zz}^{(r)}} = 1 - \frac{I_{zz}^{(r)} - I_{xx}^{(b)}}{I_{zz}^{(r)}} = 1 - \frac{I_{zz}^{(b)}\left(1 - \frac{I_{xx}^{(b)}}{I_{zz}^{(b)}}\right) + Jl_{FH}}{I_{zz}^{(b)} + Jl_{FH}} \approx 1 - \bar{l}_{FH}\frac{JR}{I_{zz}^{(b)} + Jl_{FH}},$$

where $\bar{l}_{FH} := l_{FH}/R$ is normalized flapping hinge offset. It is assumed here that the blade moments of inertia around the blade normal z_b-axis and the blade transverse x_b-axis have approximately equal values due to the high wing aspect ratio of the blade: $I_{xx}^{(b)}/I_{zz}^{(b)} \approx 1$. The ratio ϵ depends on the flapping hinge offset: if the offset is negligibly small, then this ratio is close to one; the longer is the offset, the smaller than one is this ratio. Since the impact of the flapping hinge offset on the rotor properties is analyzed in the frame of the discussed issue (see 2.6.4), the ratio is not approximated to one. The sensitivity of the ratio ϵ to the normalized hinge offset $\bar{l}_{FH}$ is represented by a coefficient ϵ_I:

$$\epsilon_I := \frac{1 - \epsilon}{\bar{l}_{FH}} \approx \frac{JR}{I_{zz}^{(b)} + Jl_{FH}} = \frac{y_{Bmc}(L_B + l_{FH})}{r_{Bz}^2 + y_{Bmc}^2 + y_{Bmc}l_{FH}}.$$

This sensitivity depends on position of the blade center of mass y_{Bmc} and on the blade radius of gyration r_{Bz} around a blade central principal axis, which is parallel to the blade normal z_b-axis and passes the blade center of mass. The blade moment of inertia $I_{zz}^{(b)} = I_{0z} + m_B y_{Bmc}^2$ is analyzed here according to the Steiner's theorem, where $I_{0z} = m_B r_{Bz}^2$ is a blade central principle moment of inertia around the central principal axis parallel to the blade normal z_b-axis passing the blade center of mass. For example, assuming a blade as a solid rod ($r_{Bz} = L_B/\sqrt{12}$) with the center of mass in the middle of the blade ($y_{Bmc} = L_B/2$) and with $\bar{l}_{FH} = 0.05$, then this sensibility equals approximately $\epsilon_I \approx 1.46$ and the ratio equals $\epsilon \approx 0.93$, that is significant decrease.

5.2.3   The simplified blade flapping equation (5.8) can be represented with the term ϵ and by dividing both sides on $\omega^2 I_{zz}^{(r)}$:

$$\epsilon \frac{1}{\omega^2} \frac{d^2\beta}{dt^2} + \beta + \frac{1}{\omega^2 I_{zz}^{(r)}} \int_0^{L_B} y_b dZ - \left((2\bar{\Omega}_x + \bar{\epsilon}_{\Omega y})\cos\psi - (2\bar{\Omega}_y - \bar{\epsilon}_{\Omega x})\sin\psi\right) = 0,$$

$$(5.9)$$

where:

$$\bar{\Omega}_x := \frac{\Omega_x}{\omega}; \quad \bar{\Omega}_y := \frac{\Omega_y}{\omega}; \quad \bar{\epsilon}_{\Omega x} := \frac{d\Omega_x}{dt}\frac{1}{\omega^2}; \quad \bar{\epsilon}_{\Omega y} := \frac{d\Omega_y}{dt}\frac{1}{\omega^2}.$$

The term $\bar{\Omega}_x$ is normalized lateral angular velocity of rotor system around the x_r-axis; the term $\bar{\Omega}_y$ is normalized longitudinal angular velocity of rotor system around the y_r-axis. Terms $\bar{\epsilon}_{\Omega x}$ and $\bar{\epsilon}_{\Omega y}$ are normalized angular accelerations of the rotor system.

The achieved equation (5.9) describes how the blade flaps under the action of the aerodynamic forces, the centrifugal forces due to rotation around the rotor shaft, the Coriolis forces due to mutual rotation of the rotor system with rotation around the rotor shaft, and the inertial forces of the accelerated rotation of the rotor system. Essentially, the equation describes blade flapping motion under the rivalry of moments of centrifugal forces with moments of aerodynamic lift forces. The moments of centrifugal forces try to eliminate blade flapping angle; the moments of aerodynamic lift forces promote the blade to increase the flapping angle in direction of action of the lift forces. This rivalry is accompanied by the blade inertial properties as a rigid mass body.

The total moment of the blade aerodynamic forces around the flapping hinge, which is represented by the third summand in the equation (5.9), depends on the blade flapping angle and blade azimuthal position; therefore, the equation is not explicit.

5.2.4   The normal component dZ of the element aerodynamic force represents the lift force of the element and can be expressed according to (3.16) in dependence on element air velocity components (3.4) and on element pitch angle (2.7), which depends on the swashplate position:

$$dZ = -\frac{\rho c(y_b) C_L^\alpha(y_b)}{2} \left[(\theta_0 + \varphi_{tw}(y_b) - \theta_1 \cos\psi - \theta_2 \sin\psi - k\beta) V_{ex}^2 + V_{ex} V_{ex} \right] dy.$$

The term $C_L^\alpha(y_b)$ is the function of element lift force coefficient slope along the blade and represents the coefficient C_{Le}^α for an element in y_b position along the blade longitudinal axis; the term $c(y_b)$ is the function of element chord versus element position and represents the element chord c_e in y_b position. The total moment of the aerodynamic forces acting on the blade around its flapping hinge is found by integrating lift force moments of all elements of the blade using the last expression with expanding of the element air velocity components V_{ex} and V_{ex} according to (3.4) and by applying the characteristic measures $B_m^{(n)}$, $\varphi_m^{(n)}$, and $\lambda^{(n,m)}(\psi)$ introduced in section 4:

$$\frac{1}{\gamma \omega^2 I_{zz}^{(r)}} \int_0^{L_B} y_b dZ = \frac{1}{\omega} \frac{d\beta}{dt} \left(\frac{B_2^{(4)}}{4} + \frac{B_2^{(3)}}{3} \mu \sin\psi \right) +$$

$$+ \beta \left[k \frac{B_1^{(4)}}{4} + \frac{B_1^{(3)}}{3} \mu (\cos\psi + 2k \sin\psi) + \frac{B_1^{(2)}}{2} \mu^2 (\sin\psi \cos\psi + k \sin^2\psi) \right] -$$

$$- \left\{ \frac{B_1^{(4)}}{4} \left(\theta_0 + \varphi_1^{(4)} \right) + 2 \frac{B_1^{(3)}}{3} \mu \left(\theta_0 + \varphi_1^{(3)} \right) \sin\psi + \frac{B_1^{(2)}}{2} \left(\theta_0 + \varphi_1^{(2)} \right) \mu^2 \sin^2\psi - \right.$$

$$- (\theta_1 \cos\psi + \theta_2 \sin\psi) \left(\frac{B_1^{(4)}}{4} + 2 \frac{B_1^{(3)}}{3} \mu \sin\psi + \frac{B_1^{(2)}}{2} \mu^2 \sin^2\psi \right) + \frac{\lambda^{(3,1)}(\psi)}{3} +$$

$$\left. + \frac{\lambda^{(2,1)}(\psi)}{2} \mu \sin\psi + (\bar{\Omega}_x \sin\psi + \bar{\Omega}_y \cos\psi) \left(\frac{B_1^{(4)}}{4} + \frac{B_1^{(3)}}{3} \mu \sin\psi \right) \right\},$$

$$\tag{5.10}$$

where

$$\gamma := \frac{\rho S_B \bar{C}_L^\alpha R^3}{2 I_{zz}^{(r)}}; \quad \mu := \frac{V_x}{\omega R}.$$

The parameter γ describes a correlation of aerodynamic and inertial properties of the blade in flapping motion performing. The heavier is the blade, the smaller is its parameter γ; the greater is the lift force, which the blade is able to create, the larger is the parameter γ. The term γ usually has values in a range of $1 - 10$. It is important to note the caveat that some literature sources use the γ coefficient with "2" in its denominator, as it is defined here; other sources define the γ coefficient without "2" in the denominator like $\gamma' := \rho S_B \bar{C}_L^\alpha R^3 / I_{zz}^{(r)}$ (the prime is used to avoid any misunderstanding). The given above definition with "2" in the denominator is used hereafter. This caveat always must be taken into account when these discussed results are compared with analogical results in other sources.

5.2.5   The derivative of blade flapping angle β with respect to time in the blade flapping equation (5.9) is useful to represent as a derivative with respect to blade azimuthal position ψ according to the relationship $\omega = d\psi/dt$:

$$\frac{d\beta}{dt} = \omega \frac{d\beta}{d\psi}.$$

The second derivative of blade flapping angle β with respect to time is similarly can be represented in terms of a second derivative with respect to blade azimuthal position ψ:

$$\frac{d^2\beta}{dt^2} = \omega^2 \frac{d^2\beta}{d\psi^2} + \frac{d\beta}{d\psi}\frac{d\omega}{dt} \approx \omega^2 \frac{d^2\beta}{d\psi^2},$$

assuming that the rotation around the rotor shaft is kept nearly constant: $\omega \approx const.$

5.2.6   Having the total moment of the blade aerodynamic forces around the flapping hinge, the blade flapping equation (5.9) is expressed in an expanded form:

$$\epsilon \frac{d^2\beta}{d\psi^2} + \gamma \frac{d\beta}{d\psi}\left(\frac{B_2^{(4)}}{4} + \frac{B_2^{(3)}}{3}\mu \sin\psi \right) +$$

$$+ \beta\left[1 + \gamma\left(k\frac{B_1^{(4)}}{4} + \frac{B_1^{(3)}}{3}\mu(\cos\psi + 2k\sin\psi) + \frac{B_1^{(2)}}{2}\mu^2(\sin\psi\cos\psi + k\sin^2\psi) \right) \right] -$$

$$- \gamma\left\{ \frac{B_1^{(4)}}{4}\left(\theta_0 + \varphi_1^{(4)} \right) + 2\frac{B_1^{(3)}}{3}\mu\left(\theta_0 + \varphi_1^{(3)} \right)\sin\psi + \frac{B_1^{(2)}}{2}\left(\theta_0 + \varphi_1^{(2)} \right)\mu^2\sin^2\psi - \right.$$

$$- (\theta_1\cos\psi + \theta_2\sin\psi)\left(\frac{B_1^{(4)}}{4} + 2\frac{B_1^{(3)}}{3}\mu\sin\psi + \frac{B_1^{(2)}}{2}\mu^2\sin^2\psi \right) + \frac{\lambda^{(3,1)}(\psi)}{3} +$$

$$+ \frac{\lambda^{(2,1)}(\psi)}{2}\mu\sin\psi + (\bar{\Omega}_x\sin\psi + \bar{\Omega}_y\cos\psi)\left(\frac{B_1^{(4)}}{4} + \frac{B_1^{(3)}}{3}\mu\sin\psi \right) \right\} -$$

$$- \left((2\bar{\Omega}_x + \bar{\varepsilon}_{\Omega y})\cos\psi - (2\bar{\Omega}_y - \bar{\varepsilon}_{\Omega x})\sin\psi \right) = 0$$

$$(5.11)$$

A solution of this second-order differential equation is the dependence of the flapping angle of the blade $\beta(\psi)$ on the blade azimuthal position ψ at given conditions and given parameters of the blade configuration. The steady rotation of the blade around the rotor shaft (see 2.4.17) is discussed here; it means that the blade repeats flapping motion at same azimuthal position every turn around the rotor shaft; therefore, the dependence $\beta(\psi)$ does not change in time. It can be concluded on the base of the repeatable blade flapping motion every turn that this dependence $\beta(\psi)$ is periodic with a period of one turn around the rotor shaft.

5.2.7   Based on approaches of the conventional blade element rotor theory, blade flapping angle versus blade azimuthal position, as a periodic solution of the flapping equation (5.11), can be represented by a Fourier series:

$$\beta(\psi) = a_0 - \sum_{k=1}^{\infty} (a_k \cos \psi + b_k \sin \psi)$$

where a_k and b_k are Fourier coefficients of a k-th harmonic, and are determined as:

$$a_k = -\frac{1}{\pi} \int_0^{2\pi} \beta(\psi) \cos(k\psi) d\psi, \quad b_k = -\frac{1}{\pi} \int_0^{2\pi} \beta(\psi) \sin(k\psi) d\psi,$$

and the Fourier constant term a_0 is determined as:

$$a_0 = \frac{1}{2\pi} \int_0^{2\pi} \beta(\psi) d\psi.$$

The Fourier coefficients remain constant in time at the steady blade rotation around the rotor shaft. In the frame of the conventional blade element rotor theory, the Fourier coefficients are interpreted based on a geometrical surface, which a blade longitudinal axis circumscribes during steady rotation around the rotor shaft. This geometrical surface has a shape of a cone with cone angle, which equals the coefficient a_0. Tilt of the axis of the cone with respect to the rotor axis is represented by the Fourier coefficients of the first harmonic ($k = 1$): the coefficient a_1 describes longitudinal tilt of the cone axis relative to the rotor frame, the coefficient b_1 describes lateral (sideways) tilt of the cone axis relative to the rotor frame (fig. 5.1). The higher harmonics ($k > 2$) represent squeezing and distortion of the blade cone surface in comparison to an ideal geometrical cone with a circular cross-section. Historically, the minus signs for harmonic Fourier coefficients were chosen to describe the positive direction of the cone axis tilt: if the a_1 coefficient is positive, then the cone axis is tilted toward azimuthal direction $\psi = 0$, i.e. backward the rotor system; if the b_1 coefficient is positive, then the cone axis is tilted toward azimuthal direction $\psi = \pi/2$, i.e. right-side the rotor system.

The Fourier coefficients decay for higher harmonics. Usually, Fourier coefficients of a certain harmonic are an order of magnitude greater than coefficients of its next harmonic. It is assumed hereafter, as well as in the conventional blade element rotor theory, that the Fourier coefficients for higher ($k > 2$) harmonics are negligibly small in comparison to the Fourier coefficients for the first harmonic ($k = 1$). Therefore, blade flapping angle versus blade azimuthal position can be approximated to

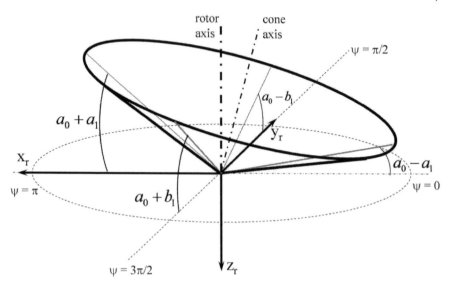

Figure 5.1 Scheme of tilted blade cone in the rotor frame with negligibly small flapping hinge offset.

the expression:

$$\beta(\psi) \approx a_0 - a_1 \cos \psi - b_1 \sin \psi. \tag{5.12}$$

With this approximation, such Fourier coefficients a_0, a_1, and b_1 should be found, which satisfies as accurate as possible the blade flapping equation (5.11). These Fourier coefficients a_0, a_1, and b_1, all together, are called blade cone parameters hereafter; a_0 is called the blade cone angle; a_1 is called the longitudinal blade cone tilt; b_1 is called the sideways (or lateral) blade cone tilt.

5.2.8 It should be admitted a feature of the rotor system with steady flapping motion approximated by first harmonics that the total moment of aerodynamic forces of a blade around its flapping hinge remains constant at no rotation of the rotor system ($\Omega_x = 0$, $\Omega_x = 0$) and with the assumption of negligibly small flapping hinge offset ($\epsilon \approx 1$). This feature can be delivered from the blade flapping equation (5.9), which transforms to the following expression with these specified conditions:

$$-\frac{1}{\omega^2 I_{zz}^{(r)}} \int_0^{L_B} y_b dZ = a_0 = const.$$

A positive element lift force is directed upward of the rotor system and is opposite to the z_r-axis: therefore $dZ < 0$. With this and with the minus in front of the left part, the cone angle a_0 is positive.

5.2.9 In order to split the blade flapping equation (5.11), the left part of this equation can be decomposed into a Fourier series. If a term $LP(\beta(\psi), \psi)$ denotes the left part of the expanded blade flapping equation (5.11), then the Fourier series of the left part can be represented as:

$$LP(\beta(\psi), \psi) = LP_0 + \sum_{k=1}^{\infty} (LP_{ak} \cos \psi + LP_{bk} \sin \psi), \quad (5.13)$$

where LP_{ak} and LP_{ak} are Fourier coefficients of a k-th harmonic:

$$LP_{ak} = \frac{1}{\pi} \int_0^{2\pi} LP(\beta(\psi), \psi) \cos(k\psi) d\psi, \quad LP_{bk} = \frac{1}{\pi} \int_0^{2\pi} LP(\beta(\psi), \psi) \sin(k\psi) d\psi,$$

and LP_0 is the Fourier constant term:

$$LP_0 = \frac{1}{2\pi} \int_0^{2\pi} LP(\beta(\psi), \psi) d\psi.$$

All these Fourier coefficients of the series must equal zero in order to satisfy the blade flapping equation (5.11). This condition is expressed as an equation system:

$$\begin{cases} LP_0 = 0 \\ ... \\ LP_{ak} = 0 \\ LP_{bk} = 0 \\ ... \end{cases},$$

where $k \in \{1, 2, ..\infty\}$.

Since it was assumed, that blade flapping is limited by first harmonic, then there are three unknown Fourier coefficients of the flapping angle: a_0, a_1, and b_1. The first three equations of the system are enough to find these coefficients:

$$\begin{cases} LP_0 = \frac{1}{2\pi} \int\limits_0^{2\pi} LP(\beta(\psi), \psi) d\psi = 0 \\ LP_{a1} = \frac{1}{\pi} \int\limits_0^{2\pi} LP(\beta(\psi), \psi) \cos \psi d\psi = 0 \\ LP_{b1} = \frac{1}{\pi} \int\limits_0^{2\pi} LP(\beta(\psi), \psi) \sin \psi d\psi = 0 \end{cases}.$$

This system is transformed into a system with three linear equations by applying the left part expression of the equation (5.11) with flapping angle β dependence on azimuthal position ψ limited by first harmonic (5.12):

$$\begin{cases} \frac{1}{\gamma}\left(1 + k\gamma\left(\frac{B_1^{(4)}}{4} + \frac{B_1^{(2)}}{4}\mu^2\right)\right)a_0 - \mu \bar{l}_{FH}\frac{B_1^{(2)}}{4}a_1 - \mu k\frac{B_1^{(3)}}{3}b_1 = c_{a0} \\ -\mu\frac{B_1^{(3)}}{3}a_0 + a_1\left(\frac{1-\epsilon}{\gamma} + k\left(\frac{B_1^{(4)}}{4} + \frac{B_1^{(2)}}{8}\mu^2\right)\right) + b_1\left(\frac{B_2^{(4)}}{4} + \frac{B_1^{(2)}}{8}\mu^2\right) = c_{b1} \\ 2k\frac{B_1^{(3)}}{3}\mu a_0 + \left(\frac{B_2^{(4)}}{4} - \frac{B_1^{(2)}}{8}\mu^2\right)a_1 - b_1\left(\frac{1-\epsilon}{\gamma} + k\left(\frac{B_1^{(4)}}{4} + 3\frac{B_1^{(2)}}{8}\mu^2\right)\right) = c_{a1} \end{cases}$$

$$(5.14)$$

where:

$$
c_{a0} = \frac{B_1^{(4)}}{4}(\theta_0 + \varphi_1^{(4)}) + \mu^2 \frac{B_1^{(2)}}{4}(\theta_0 + \varphi_1^{(2)}) - \mu \frac{B_1^{(3)}}{3}\theta_2 + \frac{\lambda_0^{(3,1)}}{3} + \mu \frac{\lambda_{b1}^{(2,1)}}{4} +
$$

$$
+ \mu \frac{B_1^{(3)}}{6}\Omega_x;
$$

$$
c_{b1} = \left(\frac{B_1^{(4)}}{4} + \frac{B_1^{(2)}}{8}\mu^2 \right)\theta_1 - \frac{\lambda_{a1}^{(3,1)}}{3} - \frac{B_1^{(4)}}{4}\Omega_y - \frac{2\Omega_x + \varepsilon_{\Omega y}}{\gamma};
$$

$$
c_{a1} = 2\mu \frac{B_1^{(3)}}{3}(\theta_0 + \varphi_1^{(3)}) + \mu \frac{\lambda_0^{(2,1)}}{2} + \frac{\lambda_{b1}^{(3,1)}}{3} - \left(\frac{B_1^{(4)}}{4} + 3\frac{B_1^{(2)}}{8}\mu^2 \right)\theta_2 +
$$

$$
+ \frac{B_1^{(4)}}{4}\Omega_x - \frac{2\Omega_y - \varepsilon_{\Omega x}}{\gamma};
$$

with assumption that azimuthal inhomogeneity of the characteristic inflow ratios (λ) are limited by only first hadronics, as was stated in 4.4.6.

A solution of the system of linear equations is a set of a_0, a_1, and b_1, which approximately describe the blade flapping motion with precision up to first harmonics of the blade flapping motion and first harmonics of the total moment of the blade aerodynamic forces versus blade azimuthal position.

5.2.10   A solution of the equation system (5.14) has complicated expressions due to the extended coefficients of the system. Introducing some coefficients simplifies the equation system to a certain extent.

A coefficient $\epsilon_\mu := B_1^{(2)}/B_2^{(4)}$ is generally used as a multiplier for $\mu^2/2$. The coefficient represents blade configuration impact on its lift force at forward motion of the rotor system. The coefficient equals one for a rectangular blade with homogeneous aerodynamic properties and negligibly short flapping hinge offset; the coefficient equals $5/3$ for a triangular blade.

The sensitivity coefficient ϵ_I (see 5.2.2) represents the flapping hinge offset impact on the flapping motion caused by different blade moments of inertia at rotation around the shaft and at flapping motion due to the flapping offset. The coefficient ϵ_I equals $(1-\epsilon)/\bar{I}_{FH}$ by definition and slightly depends on the flapping offset itself. The equation system (5.14) is rewritten in a simplified form by dividing the first and second equations of the system on $B_2^{(4)}/4$ and by applying the coefficients ϵ_I and ϵ_μ:

$$
\begin{cases}
\dfrac{1}{\gamma}\left(1 + k\gamma \left(\dfrac{B_1^{(4)}}{4} + \dfrac{B_1^{(2)}}{4}\mu^2 \right) \right)a_0 - \mu \bar{I}_{FH}\dfrac{B_1^{(2)}}{4}a_1 - \mu k \dfrac{B_1^{(3)}}{3}b_1 = c_{a0} \\[3mm]
-\mu \dfrac{4B_1^{(3)}}{3B_2^{(4)}}a_0 + a_1 \left(\bar{I}_{FH}\dfrac{4\epsilon_I}{B_2^{(4)}\gamma} + k \left(\dfrac{B_1^{(4)}}{B_2^{(4)}} + \dfrac{\epsilon_\mu \mu^2}{2} \right) \right) + b_1 \left(1 + \dfrac{\epsilon_\mu \mu^2}{2} \right) = \dfrac{4}{B_2^{(4)}}c_{b1} \\[3mm]
2k\mu \dfrac{4B_1^{(3)}}{3B_2^{(4)}}a_0 + a_1 \left(1 - \dfrac{\epsilon_\mu \mu^2}{2} \right) - b_1 \left(\bar{I}_{FH}\dfrac{4\epsilon_I}{B_2^{(4)}\gamma} + k \left(\dfrac{B_1^{(4)}}{B_2^{(4)}} + 3\dfrac{\epsilon_\mu \mu^2}{2} \right) \right) = \dfrac{4}{B_2^{(4)}}c_{a1}
\end{cases}
$$

In order to separate terms related to the impact of the flapping hinge offset, the compensation coefficient k is added and subtracted in the expressions of the multipliers for a_1 in the second equation and for b_1 in the third equation and then grouped in the following way:

$$
\begin{cases}
\dfrac{1}{\gamma}\left(1+k\gamma\left(\dfrac{B_1^{(4)}}{4}+\dfrac{B_1^{(2)}}{4}\mu^2\right)\right)a_0-\mu\bar{l}_{FH}\dfrac{B_1^{(2)}}{4}a_1-\mu k\dfrac{B_1^{(3)}}{3}b_1=c_{a0} \\[4mm]
-\mu\dfrac{4B_1^{(3)}}{3B_2^{(4)}}a_0+a_1\left(k+\bar{l}_{FH}\dfrac{4\epsilon_I}{B_2^{(4)}\gamma}+k\dfrac{B_1^{(4)}-B_2^{(4)}}{B_2^{(4)}}+k\dfrac{\epsilon_\mu\mu^2}{2}\right)+b_1\left(1+\dfrac{\epsilon_\mu\mu^2}{2}\right)= \\[4mm]
\hspace{8cm}=\dfrac{4}{B_2^{(4)}}c_{b1} \\[4mm]
2k\mu\dfrac{4B_1^{(3)}}{3B_2^{(4)}}a_0+a_1\left(1-\dfrac{\epsilon_\mu\mu^2}{2}\right)-b_1\left(k+\bar{l}_{FH}\dfrac{4\epsilon_I}{B_2^{(4)}\gamma}+k\dfrac{B_1^{(4)}-B_2^{(4)}}{B_2^{(4)}}+3k\dfrac{\epsilon_\mu\mu^2}{2}\right)= \\[4mm]
\hspace{8cm}=\dfrac{4}{B_2^{(4)}}c_{a1}
\end{cases}
$$

The blade-hinge characteristic parameter $B_2^{(4)}$ represents the non-uniform properties of the blade with the flapping hinge offset and can be expressed in terms of $B_1^{(4)}$ corrected on $\bar{l}_{FH}$ term according to the relationship (4.3):

$$
B_2^{(4)}=B_1^{(4)}-\frac{4}{3}\bar{l}_{FH}B_1^{(3)}.
$$

According to this equality, the term $(B_1^{(4)}-B_2^{(4)})/B_2^{(4)}$ directly depends on the flapping hinge offset l_{FH} and equals $\bar{l}_{FH}(4B_1^{(3)})/(3B_2^{(4)})$.

An effective compensation coefficient k_e is introduced here:

$$
k_e:=k+\frac{\bar{l}_{FH}\dfrac{4\epsilon_I}{B_2^{(4)}\gamma}+k\dfrac{B_1^{(4)}-B_2^{(4)}}{B_2^{(4)}}}{1+\dfrac{\epsilon_\mu\mu^2}{2}}=k+\bar{l}_{FH}k^{l_{FH}}, \tag{5.15}
$$

where

$$
k^{l_{FH}}:=\frac{\dfrac{4\epsilon_I}{B_2^{(4)}\gamma}}{1+\dfrac{\epsilon_\mu\mu^2}{2}}+\frac{k\dfrac{4B_1^{(3)}}{3B_2^{(4)}}}{1+\dfrac{\epsilon_\mu\mu^2}{2}}.
$$

As shown below, the effective compensation coefficient k_e represents a fictitious flapping compensation coefficient that describes the flapping compensation affected by the flapping hinge offset. This coefficient takes into account the flapping hinge offset impact by applying it into the rotor mathematical model with no flapping hinge offset for a case of hovering or vertical motion. The term $k^{l_{FH}}$ shows the sensitivity of the flapping motion to the flapping hinge offset in terms of k_e and depends on the inertial properties of the blade (the first summand) and on the fact that a blade element flaps with a different radius than a rotation radius around the rotor shaft (the second term). The equation system takes the following form with the defined k_e:

$$
\begin{cases}
\dfrac{1}{\gamma}\left(1 + k\gamma\left(\dfrac{B_1^{(4)}}{4} + \dfrac{B_1^{(2)}}{4}\mu^2\right)\right)a_0 - \mu\bar{l}_{FH}\dfrac{B_1^{(2)}}{4}a_1 - \mu k\dfrac{B_1^{(3)}}{3}b_1 = c_{a0} \\[4mm]
-\mu\dfrac{4B_1^{(3)}}{3B_2^{(4)}}a_0 + a_1 k_e\left(1 + \dfrac{\epsilon_\mu\mu^2}{2}\right) + b_1\left(1 + \dfrac{\epsilon_\mu\mu^2}{2}\right) = \dfrac{4}{B_2^{(4)}}c_{b1} \\[4mm]
2k\mu\dfrac{4B_1^{(3)}}{3B_2^{(4)}}a_0 + a_1\left(1 - \dfrac{\epsilon_\mu\mu^2}{2}\right) - b_1\left(1 + \dfrac{\epsilon_\mu\mu^2}{2}\right)\left(k_e + k\dfrac{\epsilon_\mu\mu}{1 + \dfrac{\epsilon_\mu\mu^2}{2}}\right) = \dfrac{4}{B_2^{(4)}}c_{a1}
\end{cases}
$$

$$ \tag{5.16} $$

This system of linear equations with respect to the blade cone parameters describes the blade flapping motion limited by first harmonics. This equation system is called the blade cone equation system hereafter. The solution of the cone equation system (5.16) is a set of a_0, a_1, and b_1, which is analyzed further.

5.3   SOLUTION OF THE BLADE CONE EQUATION SYSTEM

The expressions for cone parameters a_0, a_1, and b_1 of the blade flapping motion limited by first harmonics are going to be found here as the solution of the blade cone equation system in the form of (5.16). In the order of the consistent explanation, the solving is split into the following steps: firstly, the cone tilt parameters a_1 and b_1 are going to be found assuming that the parameter a_0 is known; thereafter, the parameter a_0 is going to be found based on the dependents of the cone tilt parameters a_1 and b_1 on the parameter a_0. This way is suggested for consistent introducing supplemental coefficients, which simplify the final solution expressions.

5.3.1   A linear equation system with two unknowns a_1 and b_1 is built based on the second and third equations of the blade cone equation system (5.16) assuming, that the parameter a_0 is known. The third equation of the system (5.16) represents a_1 and the second equation represents b_1 in case of no flapping compensation and with zero flapping hinge offset ($k = 0$, $l_{FH} = 0$, $k_e = 0$). The equation system for a_1 and b_1 is built in the way that a first equation represents a_1 and a second – b_1.

$$
\begin{cases}
a_1\left(1 - \dfrac{\epsilon_\mu\mu^2}{2}\right) - b_1\left(1 + \dfrac{\epsilon_\mu\mu^2}{2}\right)\left(k_e + k\dfrac{\epsilon_\mu\mu}{1 + \dfrac{\epsilon_\mu\mu^2}{2}}\right) = A_1 \\[4mm]
a_1 k_e\left(1 + \dfrac{\epsilon_\mu\mu^2}{2}\right) + b_1\left(1 + \dfrac{\epsilon_\mu\mu^2}{2}\right) = B_1
\end{cases}
, \qquad (5.17)
$$

where

$$A_1 := \frac{4}{B_2^{(4)}} c_{a1} - 2k\mu \frac{4B_1^{(3)}}{3B_2^{(4)}} a_0, \quad B_1 := \frac{4}{B_2^{(4)}} c_{b1} + \mu \frac{4B_1^{(3)}}{3B_2^{(4)}} a_0.$$

The coefficient A_1 is associated with the longitudinal blade cone tilt a_1 at the condition of no flapping compensation and with zero flapping hinge offset ($k = 0$, $k_e = 0$); the parameter B_1 is associated with the lateral blade cone tilt b_1 at the same conditions. The coefficients A_1 and B_1 depend on the blade cone angle a_0. These coefficients can be expressed with expanded c_{a1} and c_{b1}, which are introduced in (5.14):

$$A_1 = \mu \frac{8B_1^{(3)}}{3B_2^{(4)}} \left(\theta_0 - ka_0 + \varphi_1^{(3)} \right) +$$

$$+ \frac{4}{B_2^{(4)}} \left(\frac{\lambda_{b1}^{(3,1)}}{3} + \mu \frac{\lambda_0^{(2,1)}}{2} + \frac{B_1^{(4)}}{4} \bar{\Omega}_x - \frac{2\bar{\Omega}_y - \bar{\varepsilon}_{\Omega x}}{\gamma} \right) - \theta_2 \left(\frac{B_1^{(4)}}{B_2^{(4)}} + 3 \frac{\epsilon_\mu \mu^2}{2} \right);$$

$$B_1 = \mu \frac{4B_1^{(3)}}{3B_2^{(4)}} a_0 - \frac{4}{B_2^{(4)}} \left(\frac{\lambda_{a1}^{(3,1)}}{3} + \frac{B_1^{(4)}}{4} \bar{\Omega}_y + \frac{2\bar{\Omega}_x + \bar{\varepsilon}_{\Omega y}}{\gamma} \right) + \theta_1 \left(\frac{B_1^{(4)}}{B_2^{(4)}} + \frac{\epsilon_\mu \mu^2}{2} \right).$$

5.3.2 The linear equation system with respect to a_1 and b_1 can be represented in terms of linear algebra. There are a column vector $\mathbf{b}$ of dimension 2×1, which contains the unknowns a_1 and b_1, a column vector $\mathbf{B}$ of dimension 2×1, which contains parameters A_1 and B_1, and a square matrix $\mathbf{E}$ of dimension 2×2, which transforms the column vector $\mathbf{b}$ into the column vector $\mathbf{B}$ according to the linear equation system (5.17):

$$\mathbf{b} := \begin{bmatrix} a_1 \\ b_1 \end{bmatrix}; \quad \mathbf{B} := \begin{bmatrix} A_1 \\ B_1 \end{bmatrix};$$

$$\mathbf{E} := \begin{bmatrix} \left(1 - \frac{\epsilon_\mu \mu^2}{2} \right) & -\left(1 + \frac{\epsilon_\mu \mu^2}{2} \right) \left(k_e + k \frac{\epsilon_\mu \mu^2}{1 + \frac{\epsilon_\mu \mu^2}{2}} \right) \\ k_e \left(1 + \frac{\epsilon_\mu \mu^2}{2} \right) & \left(1 + \frac{\epsilon_\mu \mu^2}{2} \right) \end{bmatrix}.$$

The equation system (5.17) can be expressed in a matrix form in the following way:

$$\mathbf{Eb} = \mathbf{B}.$$

Following the linear algebra rules, the solution of this equation is found by determination of an inverse matrix of $\mathbf{E}$. The inverse matrix is denoted as a square matrix $\mathbf{D}$ of dimension 2×2 hereafter: $\mathbf{D} := \mathbf{E}^{-1}$. The column vector $\mathbf{b}$ with unknowns is found by multiplication of the inverse matrix $\mathbf{D}$ on the column vector $\mathbf{B}$:

$$\mathbf{b} = \mathbf{DB}. \tag{5.18}$$

The inverse matrix $\mathbf{D}$ is found on the base of the known elements of the matrix $\mathbf{E}$. The elements of the matrix $\mathbf{D}$ are represented in terms of supplemental coefficients, which are used hereafter:

$$\mathbf{D} = \begin{bmatrix} D_0 & D_{k2} \\ -D_k & D_{0\mu} \end{bmatrix} \tag{5.19}$$

where:

$$D_0 := \frac{1 + \dfrac{\epsilon_\mu \mu^2}{2}}{det\mathbf{E}} = \frac{1}{1 + k_e^2 - (1 - 3k_e^2)\dfrac{\epsilon_\mu \mu^2}{2} - l_{FH}\epsilon_\mu \mu^2 k_e k^{l_{FH}}},$$

$$D_{0\mu} := D_0 \left(1 - \frac{\epsilon_\mu \mu^2}{1 + \dfrac{\epsilon_\mu \mu^2}{2}} \right),$$

$$D_k := k_e D_0,$$

$$D_{k2} := D_0 \left(k_e + k \frac{\epsilon_\mu \mu^2}{1 + \dfrac{\epsilon_\mu \mu^2}{2}} \right).$$

The inverse matrix $\mathbf{D}$ can be decomposed according to cases of hovering ($\mu = 0$) and forward motion ($\mu < 0$):

$$\mathbf{D} = D_0 \left(\begin{bmatrix} 1 & k_e \\ -k_e & 1 \end{bmatrix} + \frac{\epsilon_\mu \mu^2}{1 + \dfrac{\epsilon_\mu \mu^2}{2}} \begin{bmatrix} 0 & k_e \\ 0 & -1 \end{bmatrix} + l_{FH}k^{l_{FH}} \frac{\epsilon_\mu \mu^2}{1 + \dfrac{\epsilon_\mu \mu^2}{2}} \begin{bmatrix} 0 & -1 \\ 0 & 0 \end{bmatrix} \right). \tag{5.20}$$

For example, the inverse matrix is reduced in a case of hovering ($\mu = 0$) to:

$$\mathbf{D}(\mu = 0) = D_0 \begin{bmatrix} 1 & k_e \\ -k_e & 1 \end{bmatrix}.$$

There is a relationship between the coefficients D_0, $D_{0\mu}$, D_k, and D_{k2}, which can be represented as the equality:

$$k(D_0 - D_{0\mu}) = D_{k2} - D_k.$$

Thus, the searched unknown parameters can be expressed with the introduced coefficients in the following way:

$$\begin{aligned} a_1 &= D_0 A_1 + D_{k2} B_1, \\ b_1 &= D_{0\mu} B_1 - D_k A_1. \end{aligned} \tag{5.21}$$

These cone tilt parameters a_1 and b_1 in such representation depend on the cone parameter a_0 because of the dependence of A_1 and B_1 on a_0 as shown in 5.3.1.

5.3.3   The cone parameter a_0 is found from the first equation of the cone equation system (5.16) with the found cone tilt parameters a_1 and b_1, which implicitly depend on the a_0:

$$\frac{1}{\gamma}\left(1+k\gamma\left(\frac{B_1^{(4)}}{4}+\frac{B_1^{(2)}}{4}\mu^2\right)\right)a_0-\mu\frac{B_1^{(3)}}{3}\left(\bar{l}_{FH}\frac{3B_1^{(2)}}{4B_1^{(3)}}a_1+kb_1\right)=c_{a0}. \quad (5.22)$$

Linear algebra is useful for further treatment of the expression. The column vector $\mathbf{d}_\beta$ is introduced here to collect the multipliers for a_1 and b_1 in the left part of this expression:

$$\mathbf{d}_\beta := \begin{bmatrix} \bar{l}_{FH}\dfrac{3B_1^{(2)}}{4B_1^{(3)}} \\ k \end{bmatrix}.$$

The term $\bar{l}_{FH}\dfrac{3B_1^{(2)}}{4B_1^{(3)}}a_1+kb_1$ in the right part of (5.22) is analyzed here. This expression can be represented in a matrix form as a multiplication of the transposed vector $\mathbf{d}_\beta$ on the column vector $\mathbf{b}$, which contains the cone tilt parameters a_1 and b_1:

$$\bar{l}_{FH}\frac{3B_1^{(2)}}{4B_1^{(3)}}a_1+kb_1=\mathbf{d}_\beta^T\mathbf{b}.$$

The column vector $\mathbf{b}$ was found as the solution of the equation system (5.18) in respect to a_1 and b_1 in the matrix form:

$$\bar{l}_{FH}\frac{3B_1^{(2)}}{4B_1^{(3)}}a_1+kb_1=\mathbf{d}_\beta^T\mathbf{DB}=\mathbf{D}_{a0}^T\mathbf{B},$$

where

$$\mathbf{D}_{a0} := \left(\mathbf{d}_\beta^T\mathbf{D}\right)^T.$$

The column vector $\mathbf{D}_{a0}$ collects the multipliers for A_1 and B_1, which compose the column vector $\mathbf{B}$ introduced in 5.3.2. With this representation, the analyzed expression is in terms of A_1 and B_1 instead of terms a_1 and b_1. The elements of the column vector $\mathbf{D}_{a0}$ can be found on base of the elements of the matrix $\mathbf{D}$ (5.19) and of the column vector $\mathbf{d}_\beta$. The column vector $\mathbf{D}_{a0}$ is expressed in terms of supplemental coefficients, which will be used for a_0 expression:

$$\mathbf{D}_{a0} = \begin{bmatrix} -D_{k2}^{(a0)} \\ D_k^{(a0)} \end{bmatrix}, \quad (5.23)$$

where

$$
D_k^{(a0)} := D_0 \left[k \left(1 - \frac{\epsilon \mu \mu^2}{1 + \frac{\epsilon \mu \mu^2}{2}} \right) + \bar{l}_{FH} \frac{3B_1^{(2)}}{4B_1^{(3)}} \left(k_e + k \frac{\epsilon \mu \mu^2}{1 + \frac{\epsilon \mu \mu^2}{2}} \right) \right] ;
$$

$$
D_{k2}^{(a0)} := D_0 \left(k_e k - \bar{l}_{FH} \frac{3B_1^{(2)}}{4B_1^{(3)}} \right).
$$

The coefficients $D_k^{(a0)}$ and $D_{k2}^{(a0)}$ represent the impact of the flapping compensation and the flapping hinge offset on the blade cone parameter a_0. These coefficients equal zero at no flapping compensation ($k = 0$) and at zero flapping hinge offset ($l_{FH} = 0$). These coefficients can be expressed in terms of the coefficients collected in the matrix **D**:

$$
D_k^{(a0)} = k D_{0\mu} + \bar{l}_{FH} \frac{3B_1^{(2)}}{4B_1^{(3)}} D_{k2},
$$

$$
D_{k2}^{(a0)} = k D_k - \bar{l}_{FH} \frac{3B_1^{(2)}}{4B_1^{(3)}} D_0.
$$

The analyzed term is expressed with these coefficients in the following way:

$$
\bar{l}_{FH} \frac{3B_1^{(2)}}{4B_1^{(3)}} a_1 + k b_1 = -D_{k2}^{(a0)} A_1 + D_k^{(a0)} B_1.
$$

According to the definitions of A_1 and B_1 given in (5.17), the analyzed term can be represented with the introduced column vector $\mathbf{D_{a0}}$:

$$
\bar{l}_{FH} \frac{3B_1^{(2)}}{4B_1^{(3)}} a_1 + k b_1 = \mathbf{D_{a0}^T} \begin{bmatrix} A_1 \\ B_1 \end{bmatrix} = \mathbf{D_{a0}^T} \begin{bmatrix} \dfrac{4}{B_2^{(4)}} c_{a1} - 2k\mu \dfrac{4B_1^{(3)}}{3B_2^{(4)}} a_0 \\[2ex] \dfrac{4}{B_2^{(4)}} c_{b1} + \mu \dfrac{4B_1^{(3)}}{3B_2^{(4)}} a_0 \end{bmatrix} = \tag{5.24}
$$

$$
= a_0 \mu \frac{4B_1^{(3)}}{3B_2^{(4)}} \mathbf{D_{a0}^T} \begin{bmatrix} -2k \\ 1 \end{bmatrix} + \frac{4}{B_2^{(4)}} \mathbf{D_{a0}^T} \begin{bmatrix} c_{a1} \\ c_{b1} \end{bmatrix}.
$$

5.3.4   An expression for a_0 in terms of coefficients c_{a0}, c_{a1}, and c_{b1} (5.14) is achieved by inserting the term (5.24) into the initial equation for a_0 (5.22) with following separation of the searched parameter a_0:

$$
a_0 = \gamma \frac{c_{a0} + \mu \dfrac{4B_1^{(3)}}{3B_2^{(4)}} \mathbf{D_{a0}^T} \begin{bmatrix} c_{a1} \\ c_{b1} \end{bmatrix}}{1 + k\gamma \left(\dfrac{B_1^{(4)}}{4} + \dfrac{B_1^{(2)}}{4} \mu^2 \right) - \mu^2 \gamma \dfrac{4(B_1^{(3)})^2}{9B_2^{(4)}} \mathbf{D_{a0}^T} \begin{bmatrix} -2k \\ 1 \end{bmatrix}},
$$

with the known elements of $\mathbf{D_{a0}}$ (5.23):

$$a_0 = \gamma \frac{c_{a0} - \mu \dfrac{4B_1^{(3)}}{3B_2^{(4)}} D_{k2}^{(a0)} c_{a1} + \mu \dfrac{4B_1^{(3)}}{3B_2^{(4)}} D_k^{(a0)} c_{b1}}{1 + k\gamma \left(\dfrac{B_1^{(4)}}{4} + \mu^2 \dfrac{B_1^{(2)}}{4} \right) - \gamma \mu^2 \dfrac{4(B_1^{(3)})^2}{9B_2^{(4)}} \left(D_k^{(a0)} + 2kD_{k2}^{(a0)} \right)},$$

The final expression for a_0 as a solution of the blade cone equation (5.16) is achieved by inserting the expressions for c_{a0}, c_{a1}, and c_{b1} defined in (5.14) into the last expression with following sorting of correspondent terms:

$$
\begin{aligned}
a_0 = &\left\{ \frac{B_1^{(4)}}{4}(\theta_0 + \varphi_1^{(4)}) + \mu^2 \left(\frac{B_1^{(2)}}{4}(\theta_0 + \varphi_1^{(2)}) - \frac{8(B_1^{(3)})^2}{9B_2^{(4)}} D_{k2}^{(a0)}(\theta_0 + \varphi_1^{(3)}) \right) + \right. \\
&+ \left[\frac{\lambda_0^{(3,1)}}{3} + \mu \left(\frac{\lambda_{b1}^{(2,1)}}{4} - \frac{4B_1^{(3)}}{3B_2^{(4)}} \left(D_{k2}^{(a0)} \frac{\lambda_{b1}^{(3,1)}}{3} + D_k^{(a0)} \frac{\lambda_{a1}^{(3,1)}}{3} \right) \right) - \right. \\
&\left. - \mu^2 \frac{4B_1^{(3)}}{3B_2^{(4)}} D_{k2}^{(a0)} \frac{\lambda_0^{(2,1)}}{2} \right] + \\
&+ \theta_1 \mu \frac{B_1^{(3)}}{3} D_k^{(a0)} \left[\frac{B_1^{(4)}}{B_2^{(4)}} + \frac{\epsilon \mu^2}{2} \right] - \theta_2 \mu \frac{B_1^{(3)}}{3} \left[1 - D_{k2}^{(a0)} \left(\frac{B_1^{(4)}}{B_2^{(4)}} + 3\frac{\epsilon \mu^2}{2} \right) \right] + \\
&+ \bar{\Omega}_x \mu \frac{B_1^{(3)}}{6} \left[1 - 2D_{k2}^{(a0)} \frac{B_1^{(4)}}{B_2^{(4)}} - D_k^{(a0)} \frac{16}{B_2^{(4)}\gamma} \right] - \\
&- \bar{\Omega}_y \mu \frac{B_1^{(3)}}{6} \left[2D_k^{(a0)} \frac{B_1^{(4)}}{B_2^{(4)}} - D_{k2}^{(a0)} \frac{16}{B_2^{(4)}\gamma} \right] - \\
&\left. - \mu \frac{4B_1^{(3)}}{3B_2^{(4)}} D_{k2}^{(a0)} \frac{\bar{\epsilon}_{\Omega x}}{\gamma} - \mu \frac{4B_1^{(3)}}{3B_2^{(4)}} D_k^{(a0)} \frac{\bar{\epsilon}_{\Omega y}}{\gamma} \right\} \frac{\gamma}{Denominator_{a0}},
\end{aligned}
$$

$$(5.25)$$

where

$$Denominator_{a0} := 1 + k\gamma \left(\frac{B_1^{(4)}}{4} + \mu^2 \frac{B_1^{(2)}}{4} \right) - \gamma \mu^2 \frac{4(B_1^{(3)})^2}{9B_2^{(4)}} \left(D_k^{(a0)} + 2kD_{k2}^{(a0)} \right).$$

The cone parameter a_0 basically depends on collective pitch θ_0 and on airflow passing through the rotor disc represented by a set of the characteristic inflow ratios λ. The parameter a_0 depends on the blade mass properties represented by γ: the heavier is the blade, the smaller is the cone parameter. The cone parameter a_0 as well depends on swashplate tilt (θ_1, θ_2), angular velocity ($\bar{\Omega}_x$, $\bar{\Omega}_y$), angular acceleration ($\bar{\epsilon}_{\Omega x}$, $\bar{\epsilon}_{\Omega y}$) of rotation of the whole rotor system, and azimuthal inhomogeneity of the inflow ($\lambda_{b1}^{(2,1)}$, $\lambda_{a1}^{(3,1)}$ and $\lambda_{b1}^{(3,1)}$); however, these dependencies appear at forward motion ($\mu > 0$) proportionally to μ. Dependences on $\bar{\Omega}_y$, $\bar{\epsilon}_{\Omega x}$, $\bar{\epsilon}_{\Omega y}$, $\lambda_{a1}^{(3,1)}$, $\lambda_{b1}^{(3,1)}$, and θ_1 are caused by the flapping compensation; dependences on θ_2 and $\bar{\Omega}_x$ are partially affected by the flapping compensation.

The cone tilt parameters a_1 and b_1 now can be found according to (5.21) with A_1 and B_1, which are determined with the known cone parameter a_0.

5.3.5   The introduced supplemental coefficients are discussed here in order to represent their meaning in terms of the discussed blade element rotor theory and to present their relationships with the flapping compensation and the flapping hinge offset.

The coefficient D_0 basically shows how the blade cone tilt parameters (a_1, b_1) are decreased due to flapping compensation as well as due to forward motion of the rotor system. This coefficient can be expressed with separate terms related to k and l_{FH}:

$$D_0 = \frac{1}{1 + k^2 - (1 - 3k^2)\dfrac{\epsilon_\mu \mu^2}{2} + \bar{l}_{FH} D_0^{l_{FH}}};$$

where

$$D_0^{l_{FH}} := \frac{2k(1 + \epsilon_\mu \mu^2)\left(\dfrac{4B_1^{(3)}}{3B_2^{(4)}}k + \dfrac{4\epsilon_I}{B_2^{(4)}\gamma}\right) + \bar{l}_{FH}\left(\dfrac{4B_1^{(3)}}{3B_2^{(4)}}k + \dfrac{4\epsilon_I}{B_2^{(4)}\gamma}\right)^2}{1 + \dfrac{\epsilon_\mu \mu^2}{2}}.$$

The coefficient D_0 equals one at hovering or vertical motion of the rotor system ($\mu = 0$) with no flapping compensation ($k = 0$) and with negligibly small flapping hinge offset ($l_{FH} = 0$). The coefficient is less than one with blade flapping compensation ($k > 0$); for example, $D_0 = 0.8$ for $k = 0.5$ at hovering. The $D_0^{l_{FH}}$ indicates how much D_0 is affected by the flapping hinge offset; for example, if this offset would be 5% of the rotor disk radius ($\bar{l}_{FH} = 0.05$) with $k = 0.5$, $\epsilon_I = 1.46$ and $\gamma = 4$, then $D_0 \approx 0.74$, that is on 8% less than the coefficient with negligibly small hinge offset.

The coefficient $D_{0\mu}$ shows how b_1 is affected by the flapping compensation with the flapping hinge offset and by forward motion of the rotor system:

$$D_{0\mu} = D_0\left(1 - \frac{\epsilon_\mu \mu^2}{1 + \dfrac{\epsilon_\mu \mu^2}{2}}\right).$$

This coefficient defers from D_0 at forward motion ($\mu > 0$).

The coefficient D_k shows an influence of effects in the lateral direction of the rotor system (along the y_r-axis of the rotor frame) on the cone side tilt b_1 and generally depends on the compensation coefficient. It can be rewritten in the following way:

$$D_k = kD_0 + \bar{l}_{FH} D_0 D_k^{l_{FH}},$$

where $D_k^{l_{FH}}$ indicates the influence of the flapping hinge offset on D_k,

$$D_k^{l_{FH}} := \left(\frac{4B_1^{(3)}}{3B_2^{(4)}} k + \frac{4\epsilon_I}{B_2^{(4)} \gamma} \right) \frac{1}{1 + \dfrac{\epsilon_\mu \mu^2}{2}} = k^{l_{FH}}.$$

$D_k^{l_{FH}}$ (same as $k^{l_{FH}}$) takes values in the order of magnitude of $\approx k + \epsilon_I$; for example, 5% flapping hinge offset with $\gamma = 4$ and $\epsilon_I = 1.46$ increases D_k on 12% at hovering with taking into account the implicit decrease of D_0 in the hinge offset.

A coefficient D_{k2} shows an influence of effects in the longitudinal direction of the rotor system (along the x_r-axis of the rotor frame) on cone longitudinal tilt a_1 and depends on the flapping coefficient:

$$D_{k2} = kD_0 \left(1 + \frac{\epsilon_\mu \mu^2}{1 + \dfrac{\epsilon_\mu \mu^2}{2}} \right) + \bar{l}_{FH} D_0 D_k^{l_{FH}},$$

where $D_k^{l_{FH}}$ is the same as for D_k.

The coefficients $D_k^{(a0)}$ and $D_{k2}^{(a0)}$ represent the impact of the flapping compensation and the flapping hinge offset on the cone parameter a_0 at forward motion and can be rewritten with separated terms related to k and l_{FH}:

$$D_k^{(a0)} = kD_0 \left(1 - \frac{\epsilon_\mu \mu^2}{1 + \dfrac{\epsilon_\mu \mu^2}{2}} \right) + \bar{l}_{FH} D_0 D_k^{(a0)l_{FH}};$$

$$D_{k2}^{(a0)} = k^2 D_0 + \bar{l}_{FH} D_0 D_{k2}^{(a0)l_{FH}}.$$

These coefficients are used only with the multiplier μ in the expression of a_0 (5.25) that indicates this impact appears only at forward motion proportionally μ. The $D_k^{(a0)l_{FH}}$ and $D_{k2}^{(a0)l_{FH}}$ illustrate the impact of the flapping hinge offset on the correspondent coefficients and equal:

$$D_k^{(a0)l_{FH}} := \frac{3B_1^{(2)}}{4B_1^{(3)}} \left(k + \frac{k\epsilon_\mu \mu^2 + \bar{l}_{FH} \left(\dfrac{4B_1^{(3)}}{3B_2^{(4)}} k + \dfrac{4\epsilon_I}{B_2^{(4)} \gamma} \right)}{1 + \dfrac{\epsilon_\mu \mu^2}{2}} \right);$$

$$D_{k2}^{(a0)l_{FH}} := -\frac{3B_1^{(2)}}{4B_1^{(3)}} \left(1 - \frac{4B_1^{(3)}}{3B_1^{(2)}} k \left(\frac{4B_1^{(3)}}{3B_2^{(4)}} k + \frac{4\epsilon_I}{B_2^{(4)} \gamma} \right) \frac{1}{1 + \dfrac{\epsilon_\mu \mu^2}{2}} \right).$$

The term $D_k^{(a0)l_{FH}}$ takes values in a range of the compensation coefficient k and promotes increase of $D_k^{(a0)}$; however, the term D_0 decreases with the hinge offset as shown above; so, the hinge offset with $\bar{l}_{FH} = 0.05$ cumulatively decreases $D_k^{(a0)}$ on approximately 4% in comparison with neglected offset. The term $D_{k2}^{(a0)l_{FH}}$ takes values, which can be positive or negative depending on k and ϵ_I; for example, the term $D_{k2}^{(a0)l_{FH}}$ is positive and equals 0.31 at the hinge offset with $\bar{l}_{FH} = 0.05$ with $k = 0.5$, $\epsilon_I = 1.46$, and $\gamma = 4$, and promotes increase $D_{k2}^{(a0)}$; however, the term D_0 decreases with l_{FH}, and the coefficient $D_{k2}^{(a0)}$ is cumulatively decreased on approximately 2% at hovering in comparison with neglected offset.

The denominator $Denominator_{a0}$ shows influence of the flapping compensation on the blade cone angle a_0 and can be represented in following way:

$$Denominator_{a0} =$$

$$= 1 + \frac{B_1^{(4)}}{4} k\gamma \left\{ 1 + \mu^2 \frac{B_1^{(2)}}{B_1^{(4)}} \left[1 - \frac{16}{9} \frac{(B_1^{(3)})^2}{B_1^{(2)} B_2^{(4)}} D_0 \left(1 + 2k^2 - \frac{\epsilon_\mu \mu^2}{1 + \frac{\epsilon_\mu \mu^2}{2}} \right) \right] \right\} +$$

$$+ \bar{l}_{FH} Denominator_{a0}^{l_{FH}}.$$

The flapping compensation decreases the parameter a_0; for example, if $k = 0.5$ and $\gamma = 4$ then the cone angle is decreased approximately 1.5 times in comparison with no flapping compensation at hovering ($\mu = 0$) and with rectangular blades. This decrease slightly changes with forward motion ($\mu > 0$). The $Denominator_{a0}^{l_{FH}}$ illustrates the flapping hinge offset impact on the denominator and equals:

$$Denominator_{a0}^{l_{FH}} =$$

$$= \gamma \frac{B_1^{(3)}}{3} D_0 \frac{\epsilon_\mu \mu^2}{1 + \frac{\epsilon_\mu \mu^2}{2}} \left[k \left(1 - \frac{\epsilon_\mu \mu^2}{2} \right) - \left(\frac{8B_1^{(3)}}{3B_1^{(2)}} k^2 + \bar{l}_{FH} \right) \left(\frac{4B_1^{(3)}}{3B_2^{(4)}} k + \frac{4\epsilon_I}{B_2^{(4)} \gamma} \right) \right].$$

The term $Denominator_{a0}^{l_{FH}}$ is in the order of magnitude of μ^2, which mostly is quite small value, and this term can be usually neglected.

5.3.6   It can be generally concluded within this section that the blade cone angle a_0, the longitudinal cone tilt a_1, and the lateral cone tilt b_1 can be analytically calculated according to given conditions and given rotor swashplate position/tilt.

5.4   ANALYSIS OF BLADE CONE PARAMETERS

5.4.1   Here is considered blade cone parameters of the rotor system with swashplate mechanism ($\theta_1 \neq 0$, $\theta_2 \neq 0$) and flapping compensation ($k \neq 0$) at hovering or vertical motion with the assumption of negligibly small rotation of the rotor system ($\bar{\Omega}_x = 0$, $\bar{\Omega}_y = 0$, $\bar{\varepsilon}_{\Omega x} = 0$ and $\bar{\varepsilon}_{\Omega y} = 0$). The state of hovering or vertical motion considers that forward velocity of the rotor system equals zero: $V_x = 0$, $\mu = 0$. It is assumed here that inflow through the rotor disc is similar for any azimuthal direction due to similar blade air blowing conditions at any azimuthal position; this fact ensures that the characteristic inflow ratios have no harmonics: $\lambda_{ak}^{(n,m)} = 0$, $\lambda_{bk}^{(n,m)} = 0$.

The discussed case is considered in terms of reaction of the blade cone on swashplate cyclic pitch at hovering or vertical motion.

In this case, the supplemental coefficients D_0 and $D_{0\mu}$ are the same; the coefficients D_k and D_{k2} are equal as well. These coefficients are expressed as:

$$D_0 = D_{0\mu} = \frac{1}{1+k_e^2} = \frac{1}{1+k^2+\bar{l}_{FH}D_0^{l_{FH}}},$$

$$D_k = D_{k2} = D_0 k_e = \frac{k+\bar{l}_{FH}k^{l_{FH}}}{1+k^2+\bar{l}_{FH}D_0^{l_{FH}}}.$$

Parameters A_1 and B_1 are determined only by swachplate cyclic pitch in the discussed case:

$$A_1 = -\theta_2 \frac{B_1^{(4)}}{B_2^{(4)}}, \quad B_1 = \theta_1 \frac{B_1^{(4)}}{B_2^{(4)}}.$$

Here must be admitted that the ratio $B_1^{(4)}/B_2^{(4)}$ has weak dependence on the blade non-uniform parameters and mostly depends on the flapping hinge offset; this ratio is expressed according the relationship (4.3):

$$\frac{B_1^{(4)}}{B_2^{(4)}} = \left(1 - \bar{l}_{FH}\frac{B_1^{(3)}}{B_1^{(4)}}\right)^{-1}. \tag{5.26}$$

The supplemental coefficients $D_k^{(a0)}$ and $D_{k2}^{(a0)}$ for a_0 calculation can be ignored because they are used with the multiplier μ in (5.25), which equals zero in the current case. The denominator for a_0 calculation equals $Denominator_{a0} = 1 + \frac{B_1^{(4)}}{4}k\gamma$.

Base on the stated above, the blade cone parameters can be expressed for the discussed case as:

$$a_0 = \gamma \frac{\dfrac{B_1^{(4)}}{4}(\theta_0 + \varphi_1^{(4)}) + \dfrac{\lambda_0^{(3,1)}}{3}}{1 + \dfrac{B_1^{(4)}}{4}k\gamma},$$

$$a_1 = \frac{B_1^{(4)}}{B_2^{(4)}}\frac{1}{1+k^2+\bar{l}_{FH}D_0^{l_{FH}}}\left(-\theta_2 + (k+\bar{l}_{FH}k^{l_{FH}})\theta_1\right),$$

$$b_1 = \frac{B_1^{(4)}}{B_2^{(4)}}\frac{1}{1+k^2+\bar{l}_{FH}D_0^{l_{FH}}}\left(\theta_1 + (k+\bar{l}_{FH}k^{l_{FH}})\theta_2\right).$$

Following conclusions related to the discussed case can be done according to the achieved results of the blade cone parameters:

- Flapping compensation decreases cone angle a_0. For example, for a rectangular homogeneous blade with $\gamma = 4$ and compensation coefficient $k = 0.5$ the cone angle is decreased 1.5 times in comparison with no compensation.

- Cone angle a_0 depends on the collective pitch θ_0, on the blade twisting $(\varphi_1^{(4)})$ as an additive to the collective pitch, and on the blade non-uniform properties $(B_1^{(4)})$. The greater the collective pitch, the stronger blade lift forces, the greater the cone angle.
- Tilt of the swashplate causes tilt of the blade cone. The blade cone is tilted not in the same direction than the cyclic pitch.
- If cyclic pitch has maximally negative (minimal) value in certain azimuthal direction ψ_θ with magnitude $|\theta|$ that $\theta_1 = |\theta| \cos \psi_\theta$, $\theta_2 = |\theta| \sin \psi_\theta$, then the blade cone is tilted toward azimuthal direction $\psi_{\beta=min} = \psi_\theta + \pi/2 - \arctan(k + \bar{l}_{FH} k^{l_{FH}})$. Thus, the cone tilt is ahead from the azimuthal position with the cyclic pitch minimum in direction of the shaft rotation on cone lead angle (phase lag) $\pi/2 - \arctan(k + \bar{l}_{FH} k^{l_{FH}})$, that is the same $\pi/2 - \arctan k_e$.
- The cone lead angle (phase lag) equals $\pi/2$ (90°) at no flapping compensation and negligibly small flapping offset.
- The cone lead angle (phase lag) strongly depends on the compensation coefficient but does not depend on the blade properties. For example, the lead angle equals 1.11 rad (63°) for blades with the compensation coefficient $k = 0.5$ and neglected hinge offset $\bar{l}_{FH} = 0$.
- The cone lead angle slightly depends on the flapping hinge offset. For example, the lead angle equals 1.03 rad (59°) for blades with $k = 0.5$, $\bar{l}_{FH} = 0.05$, $\epsilon_I = 1.46$, and $\gamma = 4$; thus, the 5% offset causes circa 4° decrease of the lead angle in comparison with neglected offset.
- Magnitude of the cone tilt δ_{cone} means the tilt angle of the cone axis in azimuthal direction $\psi_{\beta=min}$ of smallest flapping angle: $a_1 \cos \psi_{\beta=min} + b_1 \sin \psi_{\beta=min}$. The magnitude of cone tilt depends on the cyclic pitch magnitude $|\theta|$ and on the compensation coefficient k; it does not depend on the blade properties. At negligibly small hinge offset, the magnitude of the cone tilt is $\sqrt{1+k^2}$ times less than the cyclic pitch magnitude $|\theta|$:
$$\frac{|\theta|}{\delta_{cone}} = \sqrt{1+k^2}.$$
For example, a ratio of cone tilt magnitude to the cyclic pitch magnitude equals 0.89 for $k = 0.5$.
- Flapping hinge offset slightly changes the cone tilt magnitude in comparison with neglected offset; the cone tilt magnitude is times less then the cyclic pitch magnitude in this case:
$$\frac{|\theta|}{\delta_{cone}} = \left(1 - \bar{l}_{FH} \frac{B_1^{(3)}}{B_1^{(4)}}\right) \sqrt{1 + (k + \bar{l}_{FH} k^{l_{FH}})^2} = \left(1 - \bar{l}_{FH} \frac{B_1^{(3)}}{B_1^{(4)}}\right) \sqrt{1 + k_e^2}.$$

For example, a ratio of the cone tilt magnitude to the cyclic pitch magnitude equals 0.90 for $k = 0.5$, $\bar{l}_{FH} = 0.05$, $\epsilon_I = 1.46$, and $\gamma = 4$. Here must be noted that decrease of the ratio due to the $D_k^{l_{FH}}$ is compensated by the term $B_1^{(4)}/B_2^{(4)}$.
- The effective flapping compensation coefficient k_e represents the flapping hinge impact on the blade cone tilt. The coefficient k_e is larger than the flapping compensation coefficient k on a value proportional to the flapping hinge

offset. The effective flapping compensation coefficient can be applied into the rotor mathematical model with assumed no flapping offset in order to represent the impact at hovering and vertical flight.

5.4.2 The effect of the leading of the blade cone tilt is discussed here concerning the conditions of the previous case ($\mu = 0$, $\bar{\Omega}_x = 0$, $\bar{\Omega}_y = 0$, $\bar{\varepsilon}_{\Omega x} = 0$, and $\bar{\varepsilon}_{\Omega y} = 0$).

It is being considered here that some force cyclically affects a rotated blade. The force acts on the blade in upward direction in some blade azimuthal position, then: (i) this force creates a moment around the blade flapping hinge acting upward; (ii) the blade flaps faster upward there with rate $d\beta/dt > 0$ under this moment; (iii) this leads to decrease of attack angle of blade elements due to circumferential velocity of the flapping; (iv) that causes decrease of blade lift forces; (v) this leads to decrease of the total moment acting upward; (vi) this causes compensation of the initial force there. The opposite force affects the blade in downward direction in the opposite blade azimuthal position due to the force cyclicity, then: (i) this force creates moment around the hinge acting downward; (ii) the blade flaps faster downward there with rate $d\beta/dt < 0$; (iii) attack angle of blade elements increases there due to circumferential velocity of the flapping; (iv) that causes increase of blade lift forces; (v) this leads to decrease of the total moment acting downward; (vi) this compensates the initial force. At steady rotation, the blade flapping strives to equalize the blade force moment around the flapping hinge at all azimuthal positions by variation of the blade element attack angles caused by this flapping motion. This effect corresponds to the feature of the constant aerodynamic force moment around the flapping hinge for the rotor system with steady flapping motion limited by first harmonics, which was declared in 5.2.8. As a preliminary conclusion: the maximal upward flapping rate is in azimuthal position, where the upward force moment is expected to be highest but is eliminated by decrease of element attack angles due to this upward flapping motion; the maximal downward flapping rate is in the opposite azimuthal position, where the upward force moment is expected to be lowest but is increased by increase of attack angle due to this downward flapping motion. In an ideal case with zero flapping hinge offset and with flapping motion limited by first harmonics: the moment of the blade aerodynamic forces around the flapping hinge is constant and does not depend on blade azimuthal position; the greater this constant moment, the larger the blade cone angle a_0; the blade cyclic flapping motion (the cone tilt) is provided by the centrifugal forces and inertia of the flapping motion in this case.

According to stated above, the leading of the cone tilt as a response on swashplate cyclic pitch is explained that at steady blade rotation: (i) the maximal upward blade flapping rate is in azimuthal position, where the cyclic pitch is maximal, that causes maximal blade pitch (incidence) angle, and blade lift force is expected to be highest; (ii) the maximal downward flapping rate is in azimuthal position with minimal cyclic pitch (opposite to the azimuthal position of the previous case), that causes minimal blade pitch (incidence) angle, and the blade lift force is expected to be smallest; (iii) thus, maximal blade flapping angle, where the flapping rate is zero, is in the middle from the azimuthal position with the maximal upward flapping rate to the azimuthal position with the maximal downward flapping rate; (iv) minimal blade flapping angle consequently is in opposite azimuthal position to previous one; (v) such blade flapping motion forms a blade cone, which is tilted toward the azimuthal position with the minimal flapping angle; (vi) the cone tilts towards azimuthal position, which is ahead on $\pi/2$ in the shaft rotation direction from the azimuthal position with the minimal cyclic pitch. Thus, the blade cone responds to the cyclic pitch by tilt of the cone axis in azimuthal direction with lead angle $\pi/2$ with respect to the minimal pitch. The maximal upward flapping rate at the azimuthal position with the maximal blade pitch angle reduces blade attack angle there; the maximal downward flapping rate at azimuthal position with the minimal blade pitch angle increases the blade attack angle; thus the attack angle tries to be azimuthally homogeneous, that corresponds to the azimuthally homogeneous moment around the flapping hinge at hovering or vertical motion.

Such explanation for the $\pi/2$ lead angle is true for the rotor system with no flapping compensation ($k = 0$). The flapping compensation damps the flapping motion, which causes decrease of the lead angle (phase lag) in correspondence to the compensation coefficient: phase lag equals $\pi/2 - \arctan k$ at hovering with negligibly small flapping hinge offset ($l_{FH} \approx 0$).

5.4.3   The described effect of the blade cone transverse response on cyclic pitch is the particular case and can be generalized for any kind of impact which cyclically influences the blade flapping rate via some cyclic force. Such effects are associated in some literature as gyroscopic precession, which is inherent to any rotating system as a transverse response to a certain affect in some direction perpendicular to the rotation; however, this definition does not correspond to the aerodynamic background of the effects. In order to avoid confusions, such effects are called the cross-coupling response (of a blade cone) hereafter. According to the concept of such cross-coupling response on a certain cyclic force, a blade cone tilts toward azimuthal direction, which is ahead on lead angle from azimuthal direction, where the cyclic force maximally acts downward on the blade. The lead angle is positive and no greater than $\pi/2$. The lead angle depends on the compensation coefficient and on the flapping hinge offset. Examples of such effects can be cone response on cyclic change of blade oncoming airflow at forward motion causes the blade cone tilt backward to the forward motion, cone response on rotor system rotation ($\vec{\Omega}$) causing the cyclic attack angle change due to additional circumferential element velocity in the blade normal z-direction, that leads to the cone tilt in transverse direction relative to this rotation (as shown below).

5.4.4   In the general case, the sensitivity of the blade cone tilt on cyclic pitch caused by the swashplate tilt has a complicated dependence on the flight conditions and can be represented in extended forms based on (5.25) and (5.21):

$$\frac{da_1}{d\theta_1} = \left(\frac{B_1^{(4)}}{B_2^{(4)}} + \frac{\epsilon_\mu \mu^2}{2} \right) \left(D_{k2} + \mu^2 \gamma D_k^{(a0)} \frac{4(B_1^{(3)})^2}{9B_2^{(4)}} \frac{D_{k2} - 2kD_0}{Denominator_{a0}} \right);$$

$$\frac{da_1}{d\theta_2} = -\left(\frac{B_1^{(4)}}{B_2^{(4)}} + 3\frac{\epsilon_\mu \mu^2}{2}\right)\left(D_0 + \mu^2 \gamma D_{k2}^{(a0)} \frac{4(B_1^{(3)})^2}{9B_2^{(4)}} \frac{2D_0 k - D_{k2}}{Denominator_{a0}}\right) +$$

$$+ \mu^2 \gamma \frac{4(B_1^{(3)})^2}{9B_2^{(4)}} \frac{2D_0 k - D_{k2}}{Denominator_{a0}};$$

$$\frac{db_1}{d\theta_1} = \left(\frac{B_1^{(4)}}{B_2^{(4)}} + \frac{\epsilon_\mu \mu^2}{2}\right)\left(D_{0\mu} + \mu^2 \gamma D_k^{(a0)} \frac{4(B_1^{(3)})^2}{9B_2^{(4)}} \frac{D_{0\mu} + 2kD_k}{Denominator_{a0}}\right);$$

$$\frac{db_1}{d\theta_2} = \left(\frac{B_1^{(4)}}{B_2^{(4)}} + 3\frac{\epsilon_\mu \mu^2}{2}\right)\left(D_k + \mu^2 \gamma D_{k2}^{(a0)} \frac{4(B_1^{(3)})^2}{9B_2^{(4)}} \frac{D_{0\mu} + 2kD_k}{Denominator_{a0}}\right) -$$

$$- \mu^2 \gamma \frac{4(B_1^{(3)})^2}{9B_2^{(4)}} \frac{D_{0\mu} + 2kD_k}{Denominator_{a0}}.$$

5.4.5 In order to prove inheritance with the conventional blade element rotor theory, the following assumptions of the conventional theory are applied to the achieved above results of the cone parameters calculation.

- A blade has a rectangular shape with uniform aerodynamic properties without twisting: $c(y_b) = const$, $C_L^\alpha(y_b) = const$ and $\varphi_e(y_b) = 0$. The last assumption ensures that twisting characteristic parameters equal zero: $\varphi_m^{(n)} = 0$.
- Blade flapping hinge offset is negligibly small: $\bar{l}_{FH} \approx 0$. This assumption and the previous assumptions ensure that bade-hinge characteristic parameters equal one: $B_m^{(n)} = 1$; as well as $\epsilon_\mu = 1$. The negligibly small offset ensures as well that the ratio of moments of inertia $\epsilon = I_{xx}^{(b)}/I_{zz}^{(r)}$ approximated to one: $\epsilon \approx 1$.
- Induced velocity is homogenous around the whole rotor disk ($V_i(y_b, \psi) = const$), that ensures that the inflow ratio is constant around the rotor disk with some average value $\lambda_0 = const$: $\lambda(y_b, \psi) = \lambda_0$. In this case, the inflow ratio characteristic parameters have no harmonics ($\lambda_{ak}^{(n,m)} = 0$, $\lambda_{bk}^{(n,m)} = 0$), all average characteristic inflow ratios equal the introduced average value ($\lambda_0^{(n,m)} = \lambda_0$).
- No swashplate mechanism is used in the rotor system; no flapping compensation is applied. All blades of the rotor system always have same unchanged blade pitch angle. Thus, the cyclic pitch angles are zero ($\theta_1 = 0$ and $\theta_2 = 0$), and the compensation coefficient equals zero ($k = 0$). The constant blade pitch angle is denoted here as θ_0.
- Angular acceleration of the rotor system is assumed to be negligibly small: $\bar{\epsilon}_{\Omega x} \approx 0$, $\bar{\epsilon}_{\Omega y} \approx 0$.

With these assumptions, the supplemental coefficients for cone parameters computation are simplified:

$$D_0 = \frac{1}{1 - \frac{\mu^2}{2}}, \quad D_{0\mu} = \frac{1}{1 + \frac{\mu^2}{2}}, \quad D_k = 0, \quad \text{and} \quad D_{k2} = 0;$$

the supplemental coefficient for the a_0 calculation are zero because of $k = 0$:
$$D_k^{(a0)} = 0, \quad D_{k2}^{(a0)} = 0;$$
the denominator for a_0 equals $Denominator_{a0} = 1$.

The cone parameter a_0 is expressed in the following way with these parameters in this case:
$$a_0 = \gamma \left(\frac{1}{4} \theta_0 (1 + \mu^2) + \frac{1}{3} \lambda_0 + \frac{1}{6} \mu \bar{\Omega}_x \right).$$

The blade cone tilts are for this case:

$$a_1 = D_0 A_1 = \frac{\mu \frac{8}{3} \left(\theta_0 + \frac{3}{4} \lambda_0 \right) + \bar{\Omega}_x - \frac{8}{\gamma} \bar{\Omega}_y}{1 - \frac{\mu^2}{2}} = a_{10} + \frac{\bar{\Omega}_x - \frac{8}{\gamma} \bar{\Omega}_y}{1 - \frac{\mu^2}{2}},$$

$$b_1 = D_{0\mu} B_1 = \frac{\frac{4}{3} \mu a_0 - \bar{\Omega}_y - \frac{8}{\gamma} \bar{\Omega}_x}{1 + \frac{\mu^2}{2}} = b_{10} - \frac{\bar{\Omega}_y + \frac{8}{\gamma} \bar{\Omega}_x}{1 + \frac{\mu^2}{2}}.$$

where $a_{10} := \mu \frac{8}{3} \dfrac{\theta_0 + \frac{3}{4} \lambda_0}{1 - \frac{\mu^2}{2}}$ and $b_{10} := \dfrac{4}{3} \dfrac{\mu a_0}{1 + \frac{\mu^2}{2}}$.

These expressions correspond to the conventional blade element rotor theory.

The longitudinal cone tilt a_{10} is caused by different air blowing of blades at azimuthal positions $\pi/2$ and $3\pi/2$ at rotor system forward motion: a blade is additionally blown by the forward motion at the azimuthal position $\pi/2$ (a leading blade); a blade air blowing decreases at $3\pi/2$ (a retreating blade). The different blade air blowing causes different blade lift forces at different blade azimuthal positions: the blade lift force is stronger at $\pi/2$ and weaker in $3\pi/2$; the blade lift force changes cyclically. According to the cross-coupling response of a blade cone, the cone responds with $\pi/2$ ahead with respect to this blade lift force cyclic change; therefore the blade cone tilts toward the zero azimuthal position, which is $\pi/2$ ahead from azimuthal position with the expected minimal lift force ($\psi = 3\pi/2$) and is the opposite direction of the forward motion.

A blade, which forms a blade cone with the non-zero cone angle a_0, is differently blown along the blade normal z-axis by airflow due to forward motion: a blade in zero azimuthal position (a back blade) is blown from above; a blade in π azimuthal position (a front blade) is blown from below. This causes the different element attack angles of blades in different azimuthal positions: the attack angles are greater in π and lower in zero azimuthal position. The greater is the cone angle a_0, the greater difference of the attack angles in these azimuthal positions at forward motion. The blade lift force changes cyclically due to such attack angle variation: the lift force is expected to be stronger in π and weaker in zero azimuthal position. According to the cross-coupling response of a blade cone, the cone responds with $\pi/2$ ahead with respect to the azimuthal position with the minimal blade lift force ($\psi = 0$); therefore, the blade cone tilts toward azimuthal position $\pi/2$ due to this effect. The sideways cone tilt b_{10} represents this cone response on the different air blowing of front and back blades due to the forward motion.

5.4.6   In order to explain a response of a blade cone on angular rotation of the rotor system, here is considered a case with no flapping compensation ($k = 0$), with no swashplate mechanism ($\theta_1 = 0$, $\theta_2 = 0$), with constant blade pitch angle ($\theta_0 = const$). This case assumes rotation of the rotor system with angular velocity $\Omega_x \neq 0$, $\Omega_y \neq 0$ and angular acceleration $\bar{\varepsilon}_{\Omega x} \neq 0$, $\bar{\varepsilon}_{\Omega y} \neq 0$. The hovering or vertical motion is assumed here for simplification: $\mu = 0$. The angular velocity $\vec{\Omega}$ lays in the rotor plane along azimuthal direction ψ_Ω that $\bar{\Omega}_x = -(|\vec{\Omega}|/\omega)\cos\psi_\Omega$, $\bar{\Omega}_y = (|\vec{\Omega}|/\omega)\sin\psi_\Omega$. The circumferential velocity due to this rotation is maximal downward at $\psi_\Omega - \pi/2$ and maximal upward at $\psi_\Omega + \pi/2$. The cone tilt parameters with these conditions are expressed in the following way:

$$a_1 = -\frac{8}{B_2^{(4)}\gamma}\bar{\Omega}_y + \frac{B_1^{(4)}}{B_2^{(4)}}\bar{\Omega}_x + \frac{4}{B_2^{(4)}\gamma}\bar{\varepsilon}_{\Omega x},$$

$$b_1 = -\frac{8}{B_2^{(4)}\gamma}\bar{\Omega}_x - \frac{B_1^{(4)}}{B_2^{(4)}}\bar{\Omega}_y - \frac{4}{B_2^{(4)}\gamma}\bar{\varepsilon}_{\Omega y}.$$

The term $-\bar{\Omega}_y 8/(B_2^{(4)}\gamma)$ for a_1 and the term $-\bar{\Omega}_x 8/(B_2^{(4)}\gamma)$ for b_1 represent the impact of a moment of Coriolis forces acting on a blade due to mutual rotation around the rotor shaft accompanied by the rotation of the whole rotor system (see 3.6.6). This Coriolis force cyclically acts on the blade in presence of the rotor system rotation; this force equals zero at blade azimuthal positions $\psi_\Omega + \pi/2$ and $\psi_\Omega + 3\pi/2$; this force maximally acts downward at ψ_Ω; this force maximally acts upward at $\psi_\Omega + \pi$. This Coriolis force acts on the blade in direction, which is same as the direction of the circumferential velocity due to the rotor system rotation at blade azimuthal position on $\pi/2$ before. So, the Coriolis force is on $\pi/2$ ahead with respect to the circumferential velocity due to this rotation. According to the cross-coupling response of the blade cone, the cone responds with $\pi/2$ ahead with respect to the Coriolis force; therefore, the cone responds with π (that is $\pi/2 + \pi/2$) ahead with respect to the circumferential velocity of this rotation. It means that the cone tilts toward azimuthal direction of maximal upward circumferential velocity. This cone tilt tries to damp the rotor system rotation. The heavier is the blade, the smaller is γ, and the stronger is the cone response. This cone response depends on the non-uniform aerodynamic properties along the blade, that is represented by the parameter $B_2^{(4)}$.

The term $\bar{\Omega}_x(B_1^{(4)}/B_2^{(4)})$ for a_1 and the term $-\bar{\Omega}_y(B_1^{(4)}/B_2^{(4)})$ for b_1 represent the impact of blade element attack angle change due to the rotor system rotation. The normal z-coordinate of element air velocity (V_{ez} in 3.4) has an additional component downward at blade azimuthal position $\psi_\Omega - \pi/2$ and upward at $\psi_\Omega + \pi/2$ due to the circumferential velocity of this rotation. This causes increase of attack angles of the blade elements and increase of the total blade lift force at $\psi_\Omega - \pi/2$, decrease of the element attack angles and decrease of the blade lift force at $\psi_\Omega + \pi/2$. The blade lift force cyclically changes according to the opposite direction of this circum-ferential velocity. It can be considered as azimuthal delay of the blade lift force on π with respect to this circumferential velocity. According to the cross-coupling re-sponse of the blade cone, the cone responds with $\pi/2$ ahead with respect to the blade lift force cyclic change; therefore, the cone delays on $\pi/2$ (that is $-\pi + \pi/2$) with respect to the circumferential velocity due to rotor system rotation. It means that the blade cone tilts toward azimuthal direction, which is opposite to the vector of the angular velocity $\vec{\Omega}$ of the rotor system rotation. This response of the blade cone does not strongly depend on the blade parameters, that is represented by the mul-tiplier $B_1^{(4)}/B_2^{(4)}$, which slightly depends on the flapping hinge offset as was shown in (5.26).

The term $\bar{\varepsilon}_{\Omega x} 4/(B_2^{(4)}\gamma)$ for a_1 and the term $-\bar{\varepsilon}_{\Omega y} 4/(B_2^{(4)}\gamma)$ for b_1 represent im-pact of the inertial force due to accelerated rotation of the rotor system. These inertial forces cyclically act on a blade in the opposite direction to the accelerated circum-ferential motion of the blade due to this accelerated rotation. This can be interpreted that, the inertial forces delay on π with respect to the accelerated circumferential motion. According to the cross-coupling response of the blade cone, the cone re-sponds with $\pi/2$ ahead with respect to the inertial forces and delays on $\pi/2$ (that is $-\pi + \pi/2$) with respect to the accelerated circumferential motion. It means that the cone tilts toward azimuthal direction, which is opposite to the vector representing the accelerated rotation of the rotor system with coordinates $(d\Omega_x/dx, d\Omega_y/dx, 0)$ in the rotor frame. The heavier is the blade, the stronger is this response on accelerated rotation of the rotor system.

In a case of flapping compensation ($k \neq 0$), the lead angle of the blade cone cross-coupling response on the rotor system rotation is less than $\pi/2$ on $\arctan k$, that azimuthally shifts these responses on angle $\arctan k$ in the opposite direction of the rotor shaft rotation. Magnitude of the cone tilt, which is caused by the responses, is decreased $\sqrt{1+k^2}$ times due to the flapping compensation.

5.4.7   In the general case with flapping compensation ($k \neq 0$), with non-zero flapping hinge offset ($l_{FH} \neq 0$) and at forward motion ($\mu \neq 0$), the response of the blade cone on the rotor system rotation is easier to represent as a set of derivatives with respect to the components of this rotation $\bar{\Omega}_x$ and $\bar{\Omega}_y$. These derivatives can be found on base of the expressions for the cone tilt a_1 and b_1 (5.21) with the introduced expression for the cone angle a_0 (5.25):

$$\frac{da_1}{d\bar{\Omega}_y} = -\frac{8}{B_2^{(4)}\gamma}D_0 - \frac{B_1^{(4)}}{B_2^{(4)}}D_{k2} - \mu^2\gamma\frac{4(B_1^{(3)})^2}{9B_2^{(4)}}(D_{k2} - 2kD_0)\frac{D_k^{(a0)}\dfrac{B_1^{(4)}}{B_2^{(4)}} - D_{k2}^{(a0)}\dfrac{8}{B_2^{(4)}\gamma}}{Denominator_{a0}};$$

$$\frac{da_1}{d\bar{\Omega}_x} = \frac{B_1^{(4)}}{B_2^{(4)}}D_0 - \frac{8}{B_2^{(4)}\gamma}D_{k2}+$$

$$+\mu^2\gamma\frac{2(B_1^{(3)})^2}{9B_2^{(4)}}(D_{k2}-2kD_0)\frac{1-2D_{k2}^{(a0)}\dfrac{B_1^{(4)}}{B_2^{(4)}}-D_k^{(a0)}\dfrac{16}{B_2^{(4)}\gamma}}{Denominator_{a0}};$$

$$\frac{db_1}{d\bar{\Omega}_x} = -\frac{8}{B_2^{(4)}\gamma}D_{0\mu} - \frac{B_1^{(4)}}{B_2^{(4)}}D_k+$$

$$+\mu^2\gamma\frac{2(B_1^{(3)})^2}{9B_2^{(4)}}(D_{0\mu}+2kD_k)\frac{1-2D_{k2}^{(a0)}\dfrac{B_1^{(4)}}{B_2^{(4)}}-D_k^{(a0)}\dfrac{16}{B_2^{(4)}\gamma}}{Denominator_{a0}};$$

$$\frac{db_1}{d\bar{\Omega}_y} = -\frac{B_1^{(4)}}{B_2^{(4)}}D_{0\mu} + \frac{8}{B_2^{(4)}\gamma}D_k - \mu^2\gamma\frac{4(B_1^{(3)})^2}{9B_2^{(4)}}(D_{0\mu}+2kD_k)\frac{D_k^{(a0)}\dfrac{B_1^{(4)}}{B_2^{(4)}}-D_{k2}^{(a0)}\dfrac{8}{B_2^{(4)}\gamma}}{Denominator_{a0}}.$$

The response of the blade cone on accelerated rotation of the rotor system can be represented as a set of the derivatives of a_1 and b_1 with respect to the change rates of the coordinates of the rotor system angular velocities $\bar{\varepsilon}_{\Omega x}$ and $\bar{\varepsilon}_{\Omega y}$:

$$\frac{da_1}{d\bar{\varepsilon}_{\Omega x}} = \frac{4}{\gamma B_2^{(4)}}\left(D_0 + \mu^2\gamma\frac{4(B_1^{(3)})^2}{9B_2^{(4)}}D_{k2}^{(a0)}\frac{2kD_0-D_{k2}}{Denominator_{a0}}\right);$$

$$\frac{da_1}{d\bar{\varepsilon}_{\Omega y}} = \frac{4}{\gamma B_2^{(4)}}\left(-D_{k2} + \mu^2\gamma\frac{4(B_1^{(3)})^2}{9B_2^{(4)}}D_k^{(a0)}\frac{2kD_0-D_{k2}}{Denominator_{a0}}\right);$$

$$\frac{db_1}{d\bar{\varepsilon}_{\Omega y}} = -\frac{4}{\gamma B_2^{(4)}}\left(D_{0\mu} + \mu^2\gamma\frac{4(B_1^{(3)})^2}{9B_2^{(4)}}D_k^{(a0)}\frac{2kD_k+D_{0\mu}}{Denominator_{a0}}\right);$$

$$\frac{db_1}{d\bar{\varepsilon}_{\Omega x}} = -\frac{4}{\gamma B_2^{(4)}}\left(D_k + \mu^2\gamma\frac{4(B_1^{(3)})^2}{9B_2^{(4)}}D_{k2}^{(a0)}\frac{2kD_k+D_{0\mu}}{Denominator_{a0}}\right).$$

5.5 BLADE PITCH FORCE

5.5.1   A blade changes its pitch angle by the swashplate mechanism during the blade rotation around the rotor shaft accompanied by flapping motion. A blade pitch horn, which is solidly fixed to the blade, is swivel jointed to a vertical pitch link, which is connected to the rotating swashplate. Location of the swivel joint point is described by coordinates $\vec{p}_{pitch}(x_{pitch}^{(b)}, y_{pitch}^{(b)}, 0)$ in the blade fixed frame. During the blade rotation, the swashplate pitch link applies the blade pitch force R_{pitch} to the blade pitch horn in the joint point, which is transferred to the whole rigid blade as described in 2.4.9. This force creates a moment, which participates in blade rotation around the blade feathering hinge and ensures the blade pitch angle according to the swashplate mechanism (2.6). The blade pitch force, as well as its moment, might differ at different blade azimuthal positions depending on rotor system conditions.

5.5.2   The blade pitch force is one of the unknown parameters of the system of blade rotation equations (5.4). This parameter can be delivered from the second equation of the system with known dependence of blade flapping angle β on time:

$$
R_{pitch} = \frac{I_{yy}^{(b)} \omega^2}{x_{pitch}^{(b)}} \left(k \frac{1}{\omega^2} \frac{d^2\beta}{dt^2} - (\theta_1 \cos\psi + \theta_2 \sin\psi) - \frac{2I_{Cxx}^{(b)}}{I_{yy}^{(b)}} (\bar{\Omega}_x \sin\psi + \bar{\Omega}_y \cos\psi) + \right.
$$

$$
\left. + (\bar{\varepsilon}_{\Omega x} \cos\psi - \bar{\varepsilon}_{\Omega y} \sin\psi) - \frac{2I_{Czz}^{(b)}}{I_{yy}^{(b)}} \frac{1}{\omega} \frac{d\beta}{dt} - \frac{1}{I_{yy}^{(b)} \omega^2} \int_0^{L_B} x_{ad} dZ + \frac{1}{I_{yy}^{(b)} \omega^2} \int_0^{L_B} dM_{AD} \right).
$$

$$(5.27)$$

This expression determines which blade pitch force R_{pitch} must be applied to the blade pitch horn to satisfy the blade flapping motion $\beta(t)$ at blade azimuthal position ψ, at given conditions, and at the cyclic pitch θ_1 and θ_2 in accordance to swashplate tilt. The blade pitch force can be decomposed into two parts: a part of the force due to inertia of the rotation around the blade longitudinal y_b-axis, which is described by the first five summands in the parenthesis (5.27); and a part of the force caused by the aerodynamic forces including aerodynamic pitch moment, which is described by the last two terms in the parenthesis.

5.5.3   The part of the pitch force due to rotational inertia is cyclic for steady blade rotation around the rotor shaft. The cyclicity of this force depends on cyclic pitch θ_1 and θ_2, tilt of the blade cone, and on the rotation of the whole rotor system. The magnitude of the force part depends on the blade inertial properties around the blade longitudinal axis (I_{yy}, I_{Cxx}), on the angular speed of the shaft ω, and inversely depends on the arm of the blade pitch horn $x_{pitch}^{(b)}$. The blade pitch force can be expressed in terms of the inertial forces with the assumption of blade flapping motion limited by first harmonic (5.12) without expanding expressions for the aerodynamic moments:

$$
R_{pitch} = \frac{I_{yy}^{(b)} \omega^2}{x_{pitch}^{(b)}} \left(\frac{2I_{Czz}}{I_{yy}} b_1 + ka_1 - \theta_1 - \frac{2I_{Cxx}}{I_{yy}} \bar{\Omega}_y + \bar{\varepsilon}_{\Omega x} \right) \cos\psi +
$$

$$
+ \frac{I_{yy}^{(b)} \omega^2}{x_{pitch}^{(b)}} \left(-\frac{2I_{Czz}}{I_{yy}} a_1 + kb_1 - \theta_2 - \frac{2I_{Cxx}}{I_{yy}} \bar{\Omega}_x - \bar{\varepsilon}_{\Omega y} \right) \sin\psi -
$$

$$
- \frac{1}{x_{pitch}^{(b)}} \int_0^{L_B} x_{ad} dZ + \frac{1}{x_{pitch}^{(b)}} \int_0^{L_B} dM_{AD}.
$$

5.5.4   The part of the pitch force due to aerodynamic forces strongly depends on the element aerodynamic center position along the whole blade and on blade air blowing conditions. This part can be decomposed into a constant part and harmonics in the following way:

$$-\frac{1}{x_{pitch}^{(b)}}\int_0^{L_B} x_{ad}dZ + \frac{1}{x_{pitch}^{(b)}}\int_0^{L_B} dM_{AD} =$$

$$= \frac{I_{yy}\gamma\omega^2}{x_{pitch}^{(b)}}\left(C_{pitch0} + C_{pitchA}\cos\psi + C_{pitchB}\sin\psi + \Delta C_{pitch}\right),$$

where: C_{pitch0} is a pitch force coefficient, which represents the constant aerodynamic part; C_{pitchA} is a coefficient of a cosine first harmonic of this aerodynamic force part; C_{pitchB} is a coefficient of a sine first harmonic of this force part. These coefficients are expressed with the characteristic measures introduced in 4.6 and 4.7:

$$C_{pitch0} := \frac{B^{(3)}}{3}\left(\bar{x}_{ac}^{(3)}(\theta_0 + \varphi_{xac}^{(3)} - ka_0) + m_{ac}^{(3)}\right) +$$

$$+ \mu^2\frac{B^{(1)}}{2}\left(\bar{x}_{ac}^{(1)}(\theta_0 + \varphi_{xac}^{(1)} - ka_0) + m_{ac}^{(1)}\right) -$$

$$- \mu\bar{x}_{ac}^{(2)}\frac{B^{(2)}}{2}\left(-kb_1 + \theta_2 - a_1\bar{l}_{FH}\frac{B^{(1)}\bar{x}_{ac}^{(1)}}{B^{(2)}\bar{x}_{ac}^{(2)}} - \frac{1}{2}\bar{\Omega}_x\right) + \frac{\lambda_0^{ac(2)}}{2} + \mu\frac{\lambda_{b1}^{ac(1)}}{2};$$

$$C_{pitchA} := -\frac{B^{(3)}}{3}\bar{x}_{ac}^{(3)}\left(\theta_1 - ka_1 - b_1\left(1 - \bar{l}_{FH}\frac{3B^{(2)}\bar{x}_{ac}^{(2)}}{2B^{(3)}\bar{x}_{ac}^{(3)}}\right) - \bar{\Omega}_y\right) -$$

$$- \mu\frac{B^{(2)}}{2}\bar{x}_{ac}^{(2)}a_0 + \mu^2\frac{B^{(1)}}{4}\bar{x}_{ac}^{(1)}(b_1 + ka_1 - \theta_1) + \frac{\lambda_{a1}^{ac(2)}}{2};$$

$$C_{pitchB} := -\frac{B^{(3)}}{3}\bar{x}_{ac}^{(3)}\left(\theta_2 - kb_1 + a_1\left(1 - \bar{l}_{FH}\frac{3B^{(2)}\bar{x}_{ac}^{(2)}}{2B^{(3)}\bar{x}_{ac}^{(3)}}\right) - \bar{\Omega}_x\right) +$$

$$+ \mu B^{(2)}\left(\bar{x}_{ac}^{(2)}(\theta_0 + \varphi_{xac}^{(2)} - ka_0) + m_{ac}^{(2)}\right) - 3\mu^2\frac{B^{(1)}}{4}\bar{x}_{ac}^{(1)}\left(\theta_2 - kb_1 - \frac{1}{3}a_1\right) +$$

$$+ \frac{\lambda_{b1}^{ac(2)}}{2} + \mu\lambda_0^{ac(1)}.$$

It is assumed here that azimuthal inhomogeneity of $\lambda^{ac(n)}(\psi)$ coefficients is limited by first harmonics. The term ΔC_{pitch} represents second and third harmonics of the pitch force part due to the aerodynamic forces:

$$
\begin{aligned}
\Delta C_{pitch} = \cos 2\psi \Bigg[& \frac{B^{(2)}}{2} \mu \bar{x}_{ac}^{(2)} \left(\theta_2 - k b_1 + a_1 \left(1 - \bar{l}_{FH} \frac{B^{(1)} \bar{x}_{ac}^{(1)}}{B^{(2)} \bar{x}_{ac}^{(2)}} \right) - \frac{1}{2} \bar{\Omega}_x \right) - \\
& - \frac{B^{(1)}}{2} \mu^2 \left(\bar{x}_{ac}^{(1)} (\theta_0 + \varphi_{xac}^{(1)} - k a_0) + m_{ac}^{(1)} \right) - \mu \frac{\lambda_{b1}^{ac(1)}}{2} \Bigg] + \\
\sin 2\psi \Bigg[& - \frac{B^{(2)}}{2} \mu \bar{x}_{ac}^{(2)} \left(\theta_1 - k a_1 - b_1 \left(1 - \bar{l}_{FH} \frac{B^{(1)} \bar{x}_{ac}^{(1)}}{B^{(2)} \bar{x}_{ac}^{(2)}} \right) - \frac{1}{2} \bar{\Omega}_y \right) - \\
& - \mu^2 \frac{B^{(1)}}{2} \bar{x}_{ac}^{(1)} a_0 + \mu \frac{\lambda_{a1}^{ac(1)}}{2} \Bigg] + \\
& + \mu^2 \frac{B^{(1)}}{4} \bar{x}_{ac}^{(1)} (\theta_1 - k a_1 - b_1) \cos 3\psi + \mu^2 \frac{B^{(1)}}{4} \bar{x}_{ac}^{(1)} (\theta_2 - k b_1 + a_1) \sin 3\psi.
\end{aligned}
$$

The harmonics of the aerodynamic part of the pitch force became stronger at forward motion the rotor system; the constant part of the pitch force dominate at hovering and vertical motion.

5.5.5   According to the Newton's third law, the blade horn reacts on the swash-plate link in the swivel joint with a force $\vec{R}_{SP}$, which equals by magnitude but acts in opposite direction to the blade pitch force $\vec{R}_{pitch}$ (see 2.4.9): $\vec{R}_{SP} = -\vec{R}_{pitch}$. The force $\vec{R}_{SP}$, with which the pitch horn acts on the pitch link, transfers via the link to the swashplate and to the whole rotorcraft. This force is assumed negligibly small in comparison with a force acting on the flapping hinge and can be neglected in comparison with the total aerodynamic force created by the rotor system on the rotorcraft. The force might be important for analysis of load forces applied in a boosted control system; however, it is beyond this book. This force is neglected in further analysis.

5.6   FLAPPING HINGE REACTION MOMENT

5.6.1   The rotor shaft rotates the rotor hub together with blade flapping hinges; each flapping hinge rotates a blade, which is put on an axle of the hinge via two separate lugs (see 2.4.11). Each blade lug is affected by a hinge reaction force in a contact point with the axle; these reaction forces on the lugs create a hinge reaction moment due to separate locations of the lugs on the flapping hinge axle; this moment is directed around the normal z-axis of the blade-hinge frame. This flapping hinge reaction moment is transferred to the whole rigid blade. This moment ensures the rotation of the blade around the rotor shaft.

5.6.2   The hinge reaction moment is one of the unknown parameters of the system of blade rotation equations (5.4) and can be delivered from the third equation of the system with known dependence of blade flapping angle β on time:

$$
M_{FH} = -I_{zz}^{(r)}\frac{d\omega}{dt} - I_{zz}^{(b)}\beta\left(\frac{d\Omega_x}{dt}\cos\psi - \frac{d\Omega_y}{dt}\sin\psi\right) +
$$

$$
+ 2\omega I_{Cyy}^{(b)}\frac{d\beta}{dt}(\beta - \bar{\Omega}_x\cos\psi + \bar{\Omega}_y\sin\psi) + \int_0^{L_B} y_b dX, \tag{5.28}
$$

where $I_{zz}^{(r)} = I_{zz}^{(b)} + Jl_{FH}$ is the moment of blade inertia around the rotor axis.

A dominant component of this moment is a total moment of the aerodynamic forces of all blade elements around the normal z-axis of the blade-hinge frame, which is represented by the last summand in (5.28). This total moment of aerodynamic forces is caused by tangential components dX of element aerodynamic forces, which represent profile drag and induced drag of the blade elements. Generally, this aerodynamic moment represents a resistance moment to the blade rotation around the rotor shaft. Based on this approach, the main purpose of the hinge reaction moment is to compensate for the aerodynamic moment of the rotation resistance in order to provide the blade rotation around the rotor shaft.

The first component of the hinge reaction moment (5.28) represents inertia of the blade to its accelerated rotation around the rotor shaft. The second component is inertia of the blade to accelerated rotation of the rotor system. The third term represents the moment of Coriolis forces of rotation of the rotor system and the flapping motion accompanied by the rotation around the rotor shaft.

5.6.3   According to the Newton's third law, each blade lug affects the flapping hinge axle in its contact point with a force, which equals by magnitude and acts in opposite direction to the hinge reaction force, which affects the blade lug. The blade lug forces create the moment of blade reaction M_{onFH} acting on the flapping hinge axle due to separate locations of the lugs. This moment on the hinge axle equals by magnitude to the flapping hinge reaction moment but acts in the opposite direction $M_{onFH} = -M_{FH}$ around the normal z-axis of the blade-hinge frame. With the flapping hinge reaction moment defined above (5.28), the blade moment on the hinge equals:

$$
M_{onFH} = I_{zz}^{(r)}\frac{d\omega}{dt} + I_{zz}^{(b)}\beta\left(\frac{d\Omega_x}{dt}\cos\psi - \frac{d\Omega_y}{dt}\sin\psi\right) -
$$

$$
- 2\omega I_{Cyy}^{(b)}\frac{d\beta}{dt}(\beta - \bar{\Omega}_x\cos\psi + \bar{\Omega}_y\sin\psi) - \int_0^{L_B} y_b dX. \tag{5.29}
$$

Essentially, the blade moment on the hinge M_{onFH} transfers the aerodynamic moment around z-axis on the rotor hub.

5.6.4   The blade moment on the flapping hinge is the main part of the rotor resistance moment and will be discussed in more detail in the chapter 8 related to force moments acting on the rotor hub.

6 Force on the Flapping Hinge

The blades of the rotor system generate lift forces during the rotation around the rotor shaft. These forces are transferred through blade flapping hinges to the rotor hub and to the whole rotorcraft. The interaction of a blade with its flapping hinge in terms of this transferring of the blade forces is considered in the chapter.

6.1 FLAPPING HINGE REACTION FORCE

6.1.1 A blade has two lugs in its root part, which are put on an axle of a blade flapping hinge. The blade interacts with its flapping hinge in the contacts of the lugs with the hinge axle. Stoppers of the hinge prevent the blade lugs to slip along the axle. As stated in 2.4.11, the blade is affected in these contacts by the flapping hinge reaction force $\vec{R}_{FH}$, which is a superposition of hinge reaction forces on all blade lugs. The flapping hinge reaction force ensures the blade to be attached to the flapping hinge and ensures blade flapping motion around the axle without friction. The flapping hinge reaction force consists of the axle reaction force, which is directed perpendicularly to the axle and has just longitudinal R_{Yaxle} and normal R_{Zaxle} coordinates in the blade-hinge frame, and the reaction of the slip stoppers R_{slip} directed along the axle, the axis of which coincides with the tangential x-axis. Thus, the hinge reaction force has the coordinates $\vec{R}_{FH}(R_{slip}, R_{Yaxle}, R_{Zaxle})$ in the blade-hinge frame.

6.1.2 In order to find a flapping hinge reaction force acting on a blade, equations of motion of element center of mass (3.26) are generalized over all elements of the blade. Since the blade is considered as an absolutely rigid body, then all elements, which belong to the blade, perform same rotations and have same angular velocities. Based on this, the generalization is achieved by summing all equations of centers of mass motion of elements, which belong to the blade:

$$
\begin{aligned}
&\left(-J\frac{d^2\beta}{dt^2}\vec{e}_z + (\vec{\varepsilon}_\omega^{(h)} + \vec{\varepsilon}_\Omega^{(h)}) \times (J\vec{e}_y + m_B\vec{l}_{FH}^{(h)}) \right) + \\
&+ \left[\vec{\omega}^{(h)} \times \left(\vec{\omega}^{(h)} \times (J\vec{e}_y + m_B\vec{l}_{FH}^{(h)}) \right) + \right. \\
&\qquad \left. + \vec{\Omega}^{(h)} \times \left(\vec{\Omega}^{(h)} \times (J\vec{e}_y + m_B\vec{l}_{FH}^{(h)}) \right) - J\left(\frac{d\beta}{dt}\right)^2 \vec{e}_y \right] + \\
&+ \left\{ 2\vec{\Omega}^{(h)} \times \left(\omega^{(h)} \times (J\vec{e}_y + m_B\vec{l}_{FH}^{(h)}) - J\frac{d\beta}{dt}\vec{e}_z \right) - 2J\frac{d\beta}{dt}\vec{\omega}^{(h)} \times \vec{e}_z \right\} = \\
&= \sum_{m=1}^{Ne} d\vec{F}_{em}^{(inter)} + \sum_{m=1}^{Ne} d\vec{F}_{em}^{(exter)} - m_B\vec{A}^{(h)},
\end{aligned}
\tag{6.1}
$$

DOI: 10.1201/9781003296232-6

where m_B is the total mass of the blade and J is the mass moment of the blade; both were introduced in 5.1.1.

The sum of all internal forces between the blade elements is represented by the sum $\sum_{m=1}^{Ne} d\vec{F}_{em}^{(inter)}$, where Ne is the number of elements in the blade. This sum represents interactions between blade point particles, which do not belong to same blade element:

$$\sum_{m=1}^{Ne} d\vec{F}_{em}^{(inter)} = \sum_{m=1}^{Ne} \sum_{i \in m} \sum_{j \notin m} \vec{F}_{i,j}^{(inter)},$$

where a i-th point particle belongs to a certain m-th blade element, and a j-th point particle does not belong to the m-th blade element. This nested sum can be reordered with respect to a couple of interacting particles, each of which belongs to a different element:

$$\sum_{m=1}^{Ne} d\vec{F}_{em}^{(inter)} = \sum_{i=0}^{Np} \sum_{\substack{j=0, \\ j>i \wedge i \in m \wedge j \notin m}}^{Np} \left(\vec{F}_{i,j}^{(inter)} + \vec{F}_{j,i}^{(inter)} \right) = 0.$$

According to the Newton's third law, forces ($\vec{F}_{i,j}$ and $\vec{F}_{j,i}$) of an interacting particle couple (i and j) have same magnitude but act in opposite directions: $\vec{F}_{i,j} = -\vec{F}_{j,i}$. Therefore, the whole nested sum equal zero.

The sum of external element forces over the whole blade, which is represented by $\sum_{m=1}^{Ne} d\vec{F}_{em}^{(exter)}$, contains all external forces acting on any blade element and can be rewritten based on the definition in (3.25):

$$\sum_{m=1}^{Ne} d\vec{F}_{em}^{(exter)} = m_B \vec{g}^{(h)} + \vec{R}_{FH} + \vec{R}_{pitch} + \sum_{m=1}^{Ne} d\vec{R},$$

where: $m_B \vec{g}^{(h)}$ is the weight of the whole blade; $\sum_{m=1}^{Ne} d\vec{R}$ is a sum of element aerodynamic forces over the whole blade; $\vec{R}_{FH}$ is the flapping hinge reaction force acting on the blade lugs; $\vec{R}_{pitch}$ is the blade pitch force acting on the blade horn. As assumed in 2.5.1, the total weight of blades is much smaller than the aerodynamic forces created by the blades; therefore, the blade weight $m_B \vec{g}^{(h)}$ is neglected here. As assumed in 2.4.9 and in 5.5.5, the blade pitch force is quite small in comparison to other forces created by the rotor system; therefore, the blade pitch force is neglected here.

The inertial force due to accelerated motion of the rotor system, which is represented by the term $-m_B \vec{A}^{(h)}$ in (6.1), can be neglected based on the assumption of insignificant acceleration of the rotorcraft (see 2.5.2).

6.1.3   In order to determine each coordinate of the flapping hinge reaction force $\vec{R}_{FH}$, it is useful to decompose the generalized equation of mass center motions of all elements (6.1) into equations for each axis of the blade-hinge frame. As a result, with all of these stated above, each coordinate equation determines a correspondent coordinate of the flapping hinge reaction force:

$$
\begin{cases}
R_{slip} = -\sum_{m=1}^{Ne} dX + \dfrac{d\omega}{dt}(J\cos\beta + m_B l_{FH}) + J\sin\beta\,(\varepsilon_{\Omega x}\cos\psi - \varepsilon_{\Omega y}\sin\psi) - \\[2mm]
\quad -2J\dfrac{d\beta}{dt}\omega\sin\beta + 2J\dfrac{d\beta}{dt}\cos\beta\,(\Omega_x\cos\psi - \Omega_y\sin\psi) - \\[2mm]
\quad -(J\cos\beta + m_B l_{FH})(\Omega_x\cos\psi - \Omega_y\sin\psi)(\Omega_x\sin\psi + \Omega_y\cos\psi) \\[5mm]

R_{Yaxle} = -\sum_{m=1}^{Ne} dY - m_B l_{FH}\sin\beta\,(\varepsilon_{\Omega x}\sin\psi + \varepsilon_{\Omega y}\cos\psi) - J\left(\dfrac{d\beta}{dt}\right)^2 - \\[2mm]
\quad -\omega^2\cos\beta\,(J\cos\beta + m_B l_{FH}) + 2J\dfrac{d\beta}{dt}(\Omega_x\sin\psi + \Omega_y\cos\psi) - \\[2mm]
\quad -2\omega\sin\beta\,(J\cos\beta + m_B l_{FH})(\Omega_x\cos\psi - \Omega_y\sin\psi) + \\[2mm]
\quad +\cos\beta\,(J\cos\beta + m_B l_{FH})(\Omega_x\cos\psi - \Omega_y\sin\psi)^2 - \\[2mm]
\quad -(J + m_B l_{FH}\cos\beta)(\Omega_x^2 + \Omega_y^2) \\[5mm]

R_{Zaxle} = -\sum_{m=1}^{Ne} dZ - J\dfrac{d^2\beta}{dt^2} + (J\cos\beta + m_B l_{FH})(\varepsilon_{\Omega x}\sin\psi + \varepsilon_{\Omega y}\cos\psi) - \\[2mm]
\quad -\omega^2\sin\beta\,(J\cos\beta + m_B l_{FH}) + \\[2mm]
\quad +2\omega\cos\beta\,(J\cos\beta + m_B l_{FH})(\Omega_x\cos\psi - \Omega_y\sin\psi) + \\[2mm]
\quad +\sin\beta\,(J\cos\beta + m_B l_{FH})(\Omega_x\cos\psi - \Omega_y\sin\psi)^2 - m_B l_{FH}\sin\beta\,(\Omega_x^2 + \Omega_y^2)
\end{cases}
$$

$$(6.2)$$

Assuming that every blade element has infinitely small width ($dy_b \to 0$) that causes infinitely large number of these elements in the blade ($Ne \to \infty$), then the sums of correspondent coordinates of the element aerodynamic forces can be represented by integration:

$$
\lim_{\substack{Ne\to\infty \\ dy\to 0}}\sum_{m=1}^{Ne} dX = \int_0^{L_B} dX, \quad
\lim_{\substack{Ne\to\infty \\ dy\to 0}}\sum_{m=1}^{Ne} dY = \int_0^{L_B} dY, \quad
\lim_{\substack{Ne\to\infty \\ dy\to 0}}\sum_{m=1}^{Ne} dZ = \int_0^{L_B} dZ.
$$

6.1.4   So that, there were determined the coordinates of the flapping hinge reaction force in the blade-hinge frame (6.2), which affects the blade and ensures the blade rotation around the rotor shaft, as well as the blade flapping motion at specified conditions.

6.2   FORCE ON THE FLAPPING HINGE

6.2.1   As noticed in 2.4.11, that according to the Newton's third law, a blade acts on its flapping hinge with a force, which equals by magnitude and acts in opposite direction to the flapping hinge reaction force $\vec{R}_{FH}$, by which the flapping hinge acts on the blade through the blade lugs. The force, by which the blade acts on its flapping hinge, is called the (blade) force on the flapping hinge, is denoted $\vec{F}_{FH}$, is applied to the flapping hinge center and equals $\vec{F}_{FH} = -\vec{R}_{FH}$. The force on the hinge is represented by coordinates in the blade-hinge frame: $\vec{F}_{FH}(F_{FHx}, F_{FHy}, F_{FHz})$ as shown in fig. 6.1. The blade force on the hinge is transferred to the rotor hub and to the whole rotorcraft. The aerodynamic forces, which are created by the rotor blade and which are a part of the force on the hinge, are transferred in this way on the rotorcraft and participate in its motion.

Since the coordinates of the flapping hinge reaction force in the blade-hinge frame are determined above (6.2), then the coordinates of the force on the hinge in the same frame are known:

$$F_{FHx} = -R_{slip}, \quad F_{FHy} = -R_{Yaxle}, \quad F_{FHz} = -R_{Zaxle}.$$

6.2.2   A blade force on a flapping hinge represents instant force created by the blade at current azimuthal position and depends on current conditions. This force may change depending on blade azimuthal position at inhomogeneous azimuthal conditions of air blowing; this force is cyclically changed and repeats every turn around the rotor shaft at steady rotation and such conditions.

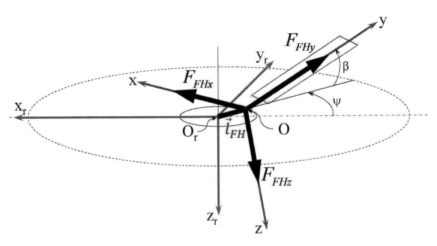

Figure 6.1   Scheme of positive coordinates of blade force on flapping hinge in blade-hinge frame.

6.2.3   It is useful for the determination of the total force of the rotor system to analyze the blade force on the flapping hinge with coordinates in the rotor frame $\vec{F}_{FH}^{(r)}(F_{FHx}^{(r)}, F_{FHy}^{(r)}, F_{FHz}^{(r)})$ as shown in fig. 6.2. This can be achieved by multiplication of a transformation matrix from the blade-hinge frame to the rotor frame $M^{(h \to r)}$ on the blade force on the flapping hinge with the coordinates in the blade-hinge frame F_{FH}. The matrix $M^{(h \to r)}$ can be found by transposition of the transformation matrix from the rotor frame to the blade-hinge frame (2.1): $M^{(h \to r)} = M^{(r \to h)T}$. With this matrix and the relationship of the force on the hinge with the hinge reaction force $(\vec{F}_{FH} = -\vec{R}_{FH})$, the force on the hinge in the rotor frame can be found as:

$$\vec{F}_{FH}^{(r)} = M^{(h \to r)} \vec{F}_{FH} = -M^{(r \to h)T} \vec{R}_{FH}.$$

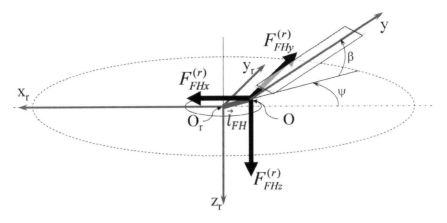

Figure 6.2 Scheme of positive coordinates of blade force on flapping hinge in rotor frame.

The coordinates of the force on the hinge in the rotor frame are found on the base of the coordinates of the hinge reaction force in the blade-hinge frame $\vec{R}_{FH}$ (6.2):

$$
\begin{aligned}
F_{FHx}^{(r)} = {} & \sin\psi \int_0^{L_B} dX - \cos\psi\cos\beta \int_0^{L_B} dY - \cos\psi\sin\beta \int_0^{L_B} dZ + \\
& + \frac{d}{dt}\left(\frac{d}{dt}\left((J\cos\beta + m_B l_{FH})\cos\psi\right) + J\sin\beta\,\Omega_y\right) + \\
& + \Omega_y\left(J\frac{d}{dt}\sin\beta - (J\cos\beta + m_B l_{FH})(\Omega_x\sin\psi + \Omega_y\cos\psi)\right)
\end{aligned}
\tag{6.3}
$$

$$
\begin{aligned}
F_{FHy}^{(r)} = {} & \cos\psi \int_0^{L_B} dX + \sin\psi\cos\beta \int_0^{L_B} dY + \sin\psi\sin\beta \int_0^{L_B} dZ + \\
& - \frac{d}{dt}\left(\frac{d}{dt}\left((J\cos\beta + m_B l_{FH})\sin\psi\right) + J\sin\beta\,\Omega_x\right) - \\
& - \Omega_x\left(J\frac{d}{dt}\sin\beta - (J\cos\beta + m_B l_{FH})(\Omega_x\sin\psi + \Omega_y\cos\psi)\right)
\end{aligned}
$$

$$
\begin{aligned}
F_{FHz}^{(r)} = {} & -\sin\beta \int_0^{L_B} dY + \cos\beta \int_0^{L_B} dZ + \\
& + \frac{d}{dt}\left(J\frac{d}{dt}\sin\beta - (J\cos\beta + m_B l_{FH})(\Omega_x\sin\psi + \Omega_y\cos\psi)\right) - \\
& - \Omega_x\left(\frac{d}{dt}\left((J\cos\beta + m_B l_{FH})\sin\psi\right) + J\sin\beta\,\Omega_x\right) - \\
& - \Omega_y\left(\frac{d}{dt}\left((J\cos\beta + m_B l_{FH})\cos\psi\right) + J\sin\beta\,\Omega_y\right)
\end{aligned}
$$

The blade aerodynamic forces, which are transferred to the blade flapping hinge, are represented here by integrals along the blade of correspondent components of element aerodynamic force (dX, dY, dZ). The components of the inertial forces are grouped after the aerodynamic components and are represented here in collapsed form, which will be useful for further purposes.

6.2.4   Thus, there were determined the components of the force (6.3), which acts by the blade on the flapping hinge at certain azimuthal position. These components are important for the determination of the total forces and moments created by the rotor system.

7 Total Rotor Force

Aerodynamic forces, which are generated by all blades of the rotor system, are transferred through the blade flapping hinges to the rotor hub and to the whole rotorcraft. These forces are generalized in the total rotor force, which represents the affection of the rotor system on its rotorcraft. The total rotor force is analyzed in this chapter.

7.1 GENERAL STATEMENTS ABOUT THE TOTAL ROTOR FORCE

7.1.1 A rotor system contains more than one blade. Each blade of the rotor system produces a force on its flapping hinge during its rotation around the rotor shaft. Each blade repeats the motion of a previous blade at steady rotation and creates same force on its flapping hinge at same azimuthal position but with some time delay. The superposition of forces created by all blades on their flapping hinges at a certain moment of time represents an instant total force created by the rotor system.

A force of each blade on its flapping hinge may periodically change every rotor turn even at steady rotation. The next blade repeats the similar periodically changed force with the time delay. In this case, all blades create vibration on the rotorcraft with frequency equal to $n\omega/(2\pi)$, where n is the blade number (see 2.4.13).

7.1.2 The blade element rotor theory operates with the averaged total force on all flapping hinges over one full shaft turn as defined in 2.4.14. This averaged force was defined as the total rotor force, which is applied to the hub center chosen as a reference point. The total rotor force is described by a vector $\vec{R}$ in the rotor frame and is decomposed into three components (fig. 7.1): the rotor thrust T is aligned along the rotor axis with positive direction upward of the rotor system (opposite to the z_r-axis of the rotor frame), the rotor longitudinal force H is directed toward zero azimuthal position, the rotor side force S is directed toward $\pi/2$ azimuthal position.

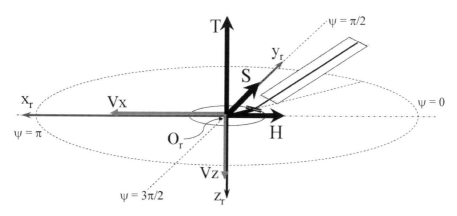

Figure 7.1 Scheme of components total rotor force in rotor frame.

DOI: 10.1201/9781003296232-7

7.1.3   Consider a rotor system, which has n flapping hinges situated on the rotor hub in axisymmetric order with attached blades. Each blade creates a force on its flapping hinge, which is described here by a vector in the rotor frame $\vec{F}^{(r)}_{FHi}(t)$. This force periodically changes at steady rotation with period $2\pi/\omega$; the subscript i denotes a certain blade out of these n blades $(i = 1..n)$. Since every blade repeats the force of its previous blade with the time delay, then a force of any blade can be described by a force of one selected blade as a reference (let it is $i = 1$) with taking into account the time delay:

$$\vec{F}^{(r)}_{FHi}(t) = \vec{F}^{(r)}_{FH1}\left(t + (i-1)\frac{Period}{n}\right),$$

where $Period = 2\pi/\omega$ is the period of one rotor turn, $Period/n$ is the time delay between two neighbur blades. An instant total force of the rotor system $\vec{R}_{instant}(t)$ is determined here as a vector sum of forces of all blades on their hinges at a certain moment of time t and can be expressed only in a term of the selected blade:

$$\vec{R}_{instant}(t) = \sum_{i=1}^{n}\vec{F}^{(r)}_{FHi}(t) = \sum_{i=1}^{n}\vec{F}^{(r)}_{FH1}\left(t + (i-1)\frac{Period}{n}\right).$$

7.1.4   The total rotor force is defined as the instant total force averaged over one rotor shaft turn and can be found by integrating the instant total force over the one period dividing on this period:

$$\vec{R} = \frac{\int_{t}^{t+Period}\vec{R}_{instant}(t)dt}{Period} = \frac{\sum_{i=1}^{n}\int_{t}^{t+Period}\vec{F}^{(r)}_{FH1}\left(t + (i-1)\frac{Period}{n}\right)dt}{Period},$$

where $\vec{R}$ is the total rotor force with coordinates in the rotor frame. Since the blade force $\vec{F}^{(r)}_{FH1}$ is periodic, then its integral over one period is always the same independently of the time delay $(i-1)\dfrac{Period}{n}$. Such integrals are the same for all blades; therefore, just an integral for one blade (for instance $i = 1$) can be used:

$$\vec{R} = \frac{n}{Period}\int_{t}^{t+Period}\vec{F}^{(r)}_{FH1}(t)dt;$$

the index of the blade is ignored further.

The components of the force on the flapping hinge is determined according to blade azimuthal position as shown in 6.2: $\vec{F}^{(r)}_{FH}(\psi)$. Therefore, it is useful to represent the integral in terms of blade azimuthal position, which corresponds to a certain moment of time. Assuming constant rotation around the rotor shaft, the variable of time t can be represented within one period by a variable of azimuthal position ψ according to the relationship $\psi = \omega t$. In this case, the integration boundaries are limited by a full round of azimuthal positions starting from convenient value of 0 and ending by 2π:

$$\vec{R} = \frac{n}{2\pi} \int\limits_{0}^{2\pi} \vec{F}_{FH}^{(r)}(\psi)\,d\psi.$$

The coordinates of the total rotor force in the rotor frame $\vec{R}(R_x, R_y, R_z)$ can be determined as averaged correspondent coordinates of the forces on all flapping hinges (6.3) over one full shaft turn:

$$R_x = \frac{n}{2\pi} \int\limits_{0}^{2\pi} \vec{F}_{FHx}^{(r)}(\psi)\,d\psi, \quad R_y = \frac{n}{2\pi} \int\limits_{0}^{2\pi} \vec{F}_{FHy}^{(r)}(\psi)\,d\psi, \quad R_z = \frac{n}{2\pi} \int\limits_{0}^{2\pi} \vec{F}_{FHz}^{(r)}(\psi)\,d\psi.$$

7.1.5   The components of the total rotor force (T, H, S) can be found based on coordinates of a blade force on a flapping hinge in the rotor frame $\vec{F}_{FH}^{(r)}(F_{FHx}^{(r)}, F_{FHy}^{(r)}, F_{FHz}^{(r)})$ determined in (6.3):

$$T = -R_z = \frac{n}{2\pi} \int\limits_{0}^{2\pi} \int\limits_{0}^{L_B} (\sin\beta\,dY - \cos\beta\,dZ)\,d\psi,$$

$$H = -R_x = \frac{n}{2\pi} \int\limits_{0}^{2\pi} \int\limits_{0}^{L_B} (-\sin\psi\,dX + \cos\psi\cos\beta\,dY + \cos\psi\sin\beta\,dZ)\,d\psi, \qquad (7.1)$$

$$S = R_y = \frac{n}{2\pi} \int\limits_{0}^{2\pi} \int\limits_{0}^{L_B} (\cos\psi\,dX + \sin\psi\cos\beta\,dY + \sin\psi\sin\beta\,dZ)\,d\psi.$$

It must be noted that averaged inertial forces do not contribute to the total rotor force at steady rotation. Integrals of the inertial forces over one shaft turn equal to zero because of equality of the following integrals to zero:

$$\int\limits_{0}^{2\pi} \frac{d^2}{dt^2}(J\cos\beta + m_B l_{FH})\cos\psi\,d\psi = 0, \quad \int\limits_{0}^{2\pi} \frac{d^2}{dt^2}(J\cos\beta + m_B l_{FH})\sin\psi\,d\psi = 0,$$

$$\int\limits_{0}^{2\pi} \frac{d}{dt}(J\cos\beta + m_B l_{FH})\cos\psi\,d\psi = 0, \quad \int\limits_{0}^{2\pi} \frac{d}{dt}(J\cos\beta + m_B l_{FH})\sin\psi\,d\psi = 0,$$

$$\int\limits_{0}^{2\pi} \frac{d}{dt}J\sin\beta\,d\psi = 0.$$

The averaged centrifugal forces due to rotor system rotation approximately equal zero because of negligibly small blade flapping angle that $\cos\beta \approx 1$:

$$\int\limits_{0}^{2\pi} (J\cos\beta + m_B l_{FH})\cos\psi \approx 0, \quad \int\limits_{0}^{2\pi} (J\cos\beta + m_B l_{FH})\sin\psi \approx 0.$$

The total rotor force consists only of the aerodynamic forces created by the blades during their rotation around the rotor shaft assuming negligibly small acceleration of the rotor system and negligibly small total weight of all blades in comparison with the force created by the rotor system. The components of the total rotor force can be found by integrating correspondent components of element aerodynamic forces (see 3.5.6).

7.2   THRUST

7.2.1   A component of the total rotor force, which is directed along the rotor axis, represents the general purpose of the rotor system to counteract the rotorcraft weight and to enable motion of the rotorcraft in the airspace. This force is originated from aerodynamic lift forces created by blades of the rotor system during rotation around the rotor shaft. This force is called the rotor thrust and is denoted T. As it was defined in 2.4.14, the rotor thrust is directed along the rotor axis with positive direction upward with respect to the rotor system, opposite to the z_r-axis of the rotor frame (fig. 7.1).

7.2.2   The rotor thrust T, as a component of the total rotor force, can be derived by integrating the expression for T (7.1) with the element aerodynamic force components determined with normalized element air velocity coordinates (3.17):

$$T = \frac{n}{2\pi} \frac{\rho(\omega R)^2}{2} \int_0^{2\pi} \int_0^{L_B} \left(c(y_b) C_L^\alpha(y_b) \cos\beta \left(\varphi_e \bar{V}_{ex}^2 + \bar{V}_{ex}\bar{V}_{ez} \right) - c(y_b) C_D(y_b) \sin\beta \bar{V}_{ex}\bar{V}_{ey} \right) dy_b d\psi,$$

where $C_L^\alpha(y_b)$ is the function of element lift force coefficient slope along a blade, $C_D(y_b)$ is the function of the element drag coefficient, $c(y_b)$ is the function of element chord on element position along a blade.

The first summand of the integrand is a projection of a blade element lift force on the rotor axis; the second is a projection of a blade element drag acting along the blade longitudinal y-axis, which appears at air blowing of the blade along its longitudinal y-axis at forward motion. The projection of such element drag is much smaller than the projection of the element lift force due to small flapping angle ($\sin\beta \ll 1$) and due to small values of an element drag coefficient in comparison to an element lift force coefficient slope ($C_D(y_b) \ll C_L^\alpha(y_b)$): therefore, $C_D(y_b)\sin\beta \ll C_L^\alpha(y_b)\cos\beta$. Based on this, the second summand of the integrand can be neglected. A constant term $S_B \bar{C}_L^\alpha$ is put out of the integral for further use of the characteristic measures introduced in section 4:

$$T = \frac{n}{2\pi} \frac{\rho S_B \bar{C}_L^\alpha (\omega R)^2}{2} \int_0^{2\pi} \int_0^{L_B} \frac{c(y_b) C_L^\alpha(y_b)}{S_B \bar{C}_L^\alpha} \cos\beta \left(\varphi_e \bar{V}_{ex}^2 + \bar{V}_{ex}\bar{V}_{ez} \right) dy_b d\psi.$$

Essentially, the total lift force of all blades is directed quite close to the axis of the blade cone, which might be tilted relative to the rotor axis according to the cyclic pitch and current conditions; therefore, the total lift force is not identical to the thrust. The thrust is a projection of this total lift force on the rotor axis, that is indicated by $\cos\beta$ in the integrand. However, the $\cos\beta$ can be approximated to one with the assumption of small flapping angle. With this, the rotor thrust can be approximated to the total lift force generated by all blades. The expression shows that the faster is the rotor shaft rotation, the stronger is the rotor thrust: $T \propto \omega^2$. The longer are the blades, the greater are circumferential velocities of the blade elements, and the stronger are the lift forces created by the blades during rotation around the rotor shaft, and the stronger is the rotor thrust: $T \propto R^4$ assuming $c_e \propto R$. The higher is the air density, the stronger are lift forces created by the blade, and the stronger is the rotor thrust: $T \propto \rho$.

7.2.3   It is established in the conventional blade element rotor theory to operate with a thrust coefficient C_T, which represents the rotor thrust normalized to the area of the rotor disc (πR^2) and to dynamic pressure of airflow on a blade tip at circumferential motion ($\rho(\omega R)^2/2$):

$$C_T := \frac{T}{\frac{1}{2}\rho\pi R^2(\omega R)^2} = \sigma \bar{C}_L^\alpha \frac{1}{2\pi}\int_0^{2\pi}\int_0^{L_B} \frac{c(y_b)C_L^\alpha(y_b)}{S_B\bar{C}_L^\alpha}\left(\varphi_e\bar{V}_{ex}^2 + \bar{V}_{ex}\bar{V}_{ez}\right)dy_b d\psi, \quad (7.2)$$

where σ is a rotor solidity, which defines a part of the rotor disc occupied by all blades of the rotor system and equals to a ratio of the total planform areas of all blades (nS_B) to the whole rotor disc area (πR^2):

$$\sigma := \frac{nS_B}{\pi R^2}.$$

The thrust coefficient is intended to generalize a rotor thrust over a whole range of size of rotor systems, rotor rotation speeds, and air densities.

7.2.4   The thrust coefficient can be determined by inserting the normalized coordinates of element air velocity (3.6) into the thrust coefficient expression (7.2) and by following integrating a received expression with applying correspondent characteristic measures introduced in the section 4:

$$C_T = \sigma \bar{C}_L^\alpha \left[\frac{B^{(3)}}{3}\left(\theta_0 + \varphi^{(3)} - ka_0\right) + \mu^2 \frac{B^{(1)}}{2}\left(\theta_0 + \varphi^{(1)} - ka_0\right) + \frac{\lambda_0^{(2)}}{2} + \right.$$
$$\left. + \mu\frac{\lambda_{b1}^{(1)}}{2} - \frac{B^{(2)}}{2}\mu\left(\theta_2 - kb_1 - a_1\bar{l}_{FH}\frac{B^{(1)}}{B^{(2)}} - \frac{1}{2}\Omega_x\right)\right]. \quad (7.3)$$

The rotor thrust depends on collective pitch θ_0: the greater is the collective pitch, the higher are pitch angles of rotor blades, the greater are lift forces produced by the blade, and the stronger is the rotor thrust.

It must be noted as a remark, that the blade element rotor theory is not able to determine the inflow ratio. Going beyond the blade element rotor theory: the stronger is a thrust produced by a rotor system, the greater is the induced velocity created by the rotor system. The induced airflow flows from the upper side of the rotor system through the rotor disc and is thrown downward when the thrust is directed upward of the rotor system. The inflow ratio is negative in this case according to its definition in 3.3.5: $\lambda < 0$. The stronger is airflow through the rotor disc downward, the greater is magnitude of the negative inflow ratio, the smaller are attack angles of blade elements due to vertical air blowing, and the weaker is the rotor thrust. The rotor system finds such necessary inflow passing through the disc $\lambda(y_b, \psi)$, which corresponds to rotor thrust at current conditions. The greatest necessary inflow ratio is at hovering and vertical motion; the necessary inflow ratio decreases at forward motion of the rotor system.

According to the expression (7.3), the thrust becomes strong with same collective pitch at forward motion ($\mu > 0$) in comparison to the hovering ($\mu = 0$) due to the second term, which indicates a higher contribution of lift forces of the advancing blades at forward flight. However, it is not a dominant reason: the necessary inflow ratio changes at forward motion and influences the thrust increase as well.

The azimuthal distribution of blade lift force is homogeneous at hovering and the vertical motion because of similar air blowing conditions of a blade at any azimuthal position. The azimuthal distribution becomes inhomogeneous at forward motion due to different air blowing of a blade in different azimuthal positions. Additionally, the induced velocity becomes inhomogeneous at forward motion; the inflow ratio acquires harmonics: $\lambda_{b1}^{(1,0)} \neq 0$. These influences on the rotor thrust at forward motion are represented by the last two summands in (7.3).

7.3 ROTOR LONGITUDINAL FORCE

7.3.1 A longitudinal force of the rotor system is a component of the total rotor force, which lays in the rotor plane and is directed opposite to the projection of the hub center velocity on the rotor plane; therefore, this force is directed toward zero azimuthal direction, that is opposite to the x_r-axis of the rotor frame (fig. 7.1). The force is denoted H. The force is applied to the hub center (the point O_r in fig. 7.1), which is specified as a reference point in 2.4.14. The longitudinal force is much smaller than the rotor thrust. However, the main purpose of this force is ability to be changed in accordance with tilt of the rotor swashplate and to create a lateral moment about the rotorcraft center of mass, which enables control over the rotorcraft.

7.3.2 The longitudinal rotor force can be derived by integrating the expression of H (7.1) with applying the element aerodynamic force components (3.17). The force can be split into two parts: a part caused by blade lift forces; and a part caused by a blade profile drag:

$$H = \frac{\rho n S_B \bar{C}_L^\alpha (\omega R)^2}{2} \left[-\frac{1}{2\pi} \int_0^{2\pi} \int_0^{L_B} \frac{c(y_b) C_L^\alpha(y_b)}{S_B \bar{C}_L^\alpha} (\varphi_e \bar{V}_{ex} + \bar{V}_{ez}) \times \right.$$

$$\left. \times (\bar{V}_{ex} \sin\beta \cos\psi + \bar{V}_{ez} \sin\psi) \, dy_b d\psi \right] +$$

$$+ \frac{\rho n S_B (\omega R)^2}{2} \frac{1}{2\pi} \int_0^{2\pi} \int_0^{L_B} \frac{c(y_b) C_D(y_b)}{S_B} \bar{V}_{ex} (\bar{V}_{ex} \sin\psi - \bar{V}_{ey} \cos\psi \cos\beta) \, dy_b d\psi.$$

(7.4)

7.3.3 Analogous to the thrust coefficient, a longitudinal force coefficient C_H is introduced in order to generalize the longitudinal force over the rotor system size, the rotor shaft angular speed, and air density. The coefficient C_H is defined as the longitudinal force normalized to the area of the rotor disc (πR^2) and to dynamic pressure of airflow on a blade tip at circumferential motion $(\rho(\omega R)^2/2)$:

$$C_H := \frac{H}{\frac{1}{2}\rho\pi R^2(\omega R)^2} = \sigma\bar{C}_L^\alpha \left[-\frac{1}{2\pi}\int_0^{2\pi}\int_0^{L_B}\frac{c(y_b)C_L^\alpha(y_b)}{S_B\bar{C}_L^\alpha}(\varphi_e\bar{V}_{ex}+\bar{V}_{ez}) \times \right.$$

$$\left. \times (\bar{V}_{ex}\sin\beta\cos\psi+\bar{V}_{ez}\sin\psi)\,dy_b\,d\psi \right] + \tag{7.5}$$

$$+\sigma\frac{1}{2\pi}\int_0^{2\pi}\int_0^{L_B}\frac{c(y_b)C_D(y_b)}{S_B}\bar{V}_{ex}(\bar{V}_{ex}\sin\psi-\bar{V}_{ey}\cos\beta\cos\psi)\,dy_b\,d\psi.$$

7.3.4 The term $(\bar{V}_{ex}\cos\psi\sin\beta+\bar{V}_{ez}\sin\psi)$ is analyzed here in order to explain the structure of the part of the longitudinal force, which is caused by projections of blade lift forces on the rotor longitudinal direction (the x_r-axis) and is represented by the first summand in (7.5). The term can be rewritten in terms of the normalized coordinates of element air velocity (3.6), with the assumption of the small blade flapping angle, that $\sin\beta\approx\beta$ and $\cos\beta\approx 1$, and with flapping motion limited by first harmonics (5.12). The summands of an achieved expression can be reordered in the following way:

$$\bar{V}_{ex}\beta\cos\psi+\bar{V}_{ez}\sin\psi = -a_1\bar{V}_{ex}+a_0\bar{V}_{ex}\cos\psi+$$

$$-\sin\psi\left(\mu(a_0\cos\psi-a_1)-\lambda(y_b,\psi)-\frac{l_{FH}+y_b}{R}(\bar{\Omega}_x\sin\psi+\bar{\Omega}_y\cos\psi)\right)+$$

$$+\bar{l}_{FH}(a_1\sin\psi-b_1\cos\psi)\sin\psi.$$

Inserting this expression into the expression of C_H, and splitting an achieved expression according to these summands:

$$C_H = a_1\left[\frac{\sigma\bar{C}_L^\alpha}{2\pi}\int_0^{2\pi}\int_0^{L_B}\frac{c(y_b)C_L^\alpha(y_b)}{S_B\bar{C}_L^\alpha}(\varphi_e\bar{V}_{ex}+\bar{V}_{ez})\bar{V}_{ex}\,dy_b\,d\psi\right]-$$

$$-a_0\left[\frac{\sigma\bar{C}_L^\alpha}{2\pi}\int_0^{2\pi}\int_0^{L_B}\frac{c(y_b)C_L^\alpha(y_b)}{S_B\bar{C}_L^\alpha}(\varphi_e\bar{V}_{ex}+\bar{V}_{ez})\bar{V}_{ex}\cos\psi\,dy_b\,d\psi\right]+$$

$$+\left[\frac{\sigma\bar{C}_L^\alpha}{2\pi}\int_0^{2\pi}\int_0^{L_B}\frac{c(y_b)C_L^\alpha(y_b)}{S_B\bar{C}_L^\alpha}(\varphi_e\bar{V}_{ex}+\bar{V}_{ez})\sin\psi\times\right.$$

$$\left. \times\left(\mu(a_0\cos\psi-a_1)-\lambda(y_b,\psi)-\frac{l_{FH}+y_b}{R}(\bar{\Omega}_x\sin\psi+\bar{\Omega}_y\cos\psi)\right)dy_b\,d\psi\right]- \tag{7.6}$$

$$-\bar{l}_{FH}\left[\frac{\sigma\bar{C}_L^\alpha}{2\pi}\int_0^{2\pi}\int_0^{L_B}\frac{c(y_b)C_L^\alpha(y_b)}{S_B\bar{C}_L^\alpha}(\varphi_e\bar{V}_{ex}+\bar{V}_{ez})\sin\psi(a_1\sin\psi-b_1\cos\psi)\,dy_b\,d\psi\right]+$$

$$+\frac{\sigma}{2\pi}\int_0^{2\pi}\int_0^{L_B}\frac{c(y_b)C_D(y_b)}{S_B}\bar{V}_{ex}(\bar{V}_{ex}\sin\psi-\bar{V}_{ey}\cos\psi)\,dy_b\,d\psi.$$

In a view of such presentation of the longitudinal force, it consists of following components:

- The expression in square brackets of the first summand represents the total lift force of all blades. The total lift force is closely directed to the axis of the blade cone: the total lift force tilts together with the blade cone. If the blade cone has some longitudinal tilt, then the total lift force has a projection on the rotor longitudinal direction. The multiplier a_1 indicates the projection of the total lift force on the zero azimuthal direction. The total lift force approximately equals the rotor thrust due to small flapping angle, as was shown in 7.2.2. So that, the first summand equals $a_1 C_T$ according to the thrust coefficient definition (7.2). The greater is blade cone tilt in the longitudinal direction with angle a_1, the stronger is the rotor longitudinal force.

- Basically, a lift force of a blade element is aligned almost along the normal z-axis of the blade-hinge frame with slight tilt in opposite direction to the element velocity; however, this tilt is neglected in the current consideration. Forming a cone with cone angle a_0, the blade lift force is not perpendicular to the rotor plane due to blade flapping: the lift force has non-zero projection on the rotor plane in radial direction to the hub center. If the blade lift force is azimuthally homogeneous, then the radial projections of the lift forces are compensated, and an averaged projection of the lift forces due to the blade cone equals zero; if the blade lift force in the front part of the rotor disc differs from lift force in the back part, then radial projections of the forces on the rotor plane are not compensated and force difference acts in the rotor longitudinal direction. This effect is represented by the second summand of the rotor longitudinal force; the multiplier $\cos \psi$ of its integrand indicates importance of the lift force inhomogeneity in the rotor longitudinal direction. Such longitudinal inhomogeneity of the lift force usually appears at forward motion ($\mu > 0$), when a front blade ($\psi = \pi$) is blown from below by oncoming airflow due to the rotor system motion and a back blade ($\psi = 0$) is blown from above. The larger is the cone angle a_0 at inhomogeneous lift forces in rotor longitudinal direction, the stronger is this component: the a_0 multiplier in the second summand indicates this.

- A lift force of a blade element is aligned along the normal z-axis with slight tilt in opposite direction to the element velocity, which causes the induced element drag described in 3.5.7. The element induced drag acts in the tangential direction parallel to the rotor plane: the induced drag is directed backward of the rotor system for a blade in azimuthal position $\pi/2$ and forward for a blade in azimuthal position $3\pi/2$ (at no autorotation). If the air blowing conditions are azimuthally homogeneous, then the induced drag forces are compensated and the averaged induced drag equals zero; if there is lateral inhomogeneity of the air blowing condition, then the difference of the induced drag forces of advancing and retreating blades creates an additional force component in the rotor longitudinal direction. Such force is represented by the third summand, where the multiplier $\sin \psi$ in the integrand indicates importance of the lateral inhomogeneity of the induced drag. Such lateral inhomogeneity appears at forward motion as well as at lateral rotation of the rotor system. Additionally, azimuthal inhomogeneity of induced airflow in the rotor lateral direction appears at forward motion, that influences this component as well.

- The previous summands are determined by blade flapping motion assuming negligibly small flapping hinge offset. The fourth summand represents a compensation force term caused by the non-flapping part of the hinge offset.
- The fifth summand represents the averaged profile drag of all elements of all blades projected on the zero azimuthal direction. A blade profile drag acts backward for an advancing blade at azimuthal position $\pi/2$ and forward for a retreating blade at azimuthal position $3\pi/2$. These forces are equal by magnitude and act in opposite direction at hovering and vertical motion; therefore, they are compensated and the profile drag does not contribute to the longitudinal rotor force. At forward motion, the advancing blade has higher air velocity and higher profile drag; the retreading blade has lower air velocity and lower profile drag; there is a difference in the profile drag forces that contributes to the longitudinal rotor force. At forward motion, the front blade ($\psi = \pi$) is additionally blown along the blade longitudinal axis toward the flapping hinge and affected by the profile drag, which is directed backward to the rotor system; the back blade ($\psi = 0$) is as well blown along the blade longitudinal axis but toward the blade tip and is affected by the profile drag also directed backward to the rotor system; this contributes to the rotor longitudinal force.

7.3.5   A final expression of the longitudinal rotor force coefficient can be achieved by integrating all summands (7.6) with expanded normalized components of element air velocities according to (3.6), with the characteristic measures introduced in 4, and with following component sorting:

$$
\begin{aligned}
C_H = {} & a_1 C_T + \frac{3}{4}\mu\sigma C_D^{(2)} + \\
& + \sigma \bar{C}_L^{\alpha}\Bigg\{ \mu\frac{B^{(2)}}{4}a_0^2 - \mu^2\frac{B^{(1)}}{2}a_1\left(\theta_0 + \varphi^{(1)} - ka_0\right) + \\
& \quad + (\theta_1 - b_1 - ka_1)\left[\frac{B^{(3)}}{6}a_0 + \mu\frac{\lambda_{a1}^{(1)}}{8}\right] + \\
& \quad + (\theta_2 + a_1 - kb_1)\left[\mu\frac{B^{(2)}}{4}a_1 + \frac{\lambda_0^{(2)}}{4} + 3\mu\frac{\lambda_{b1}^{(1)}}{8}\right] - \\
& \quad - \bar{\Omega}_x\left[\frac{B^{(3)}}{6}\left(\theta_0 + \varphi^{(3)} - ka_0\right) - 3\mu\frac{B^{(2)}}{16}\left(\theta_2 - \frac{5}{3}a_1 - kb_1\right) + \frac{\lambda_0^{(2)}}{2}\right] \\
& \quad - \bar{\Omega}_y\left[\frac{B^{(3)}}{6}a_0 - \mu\frac{B^{(2)}}{16}(\theta_1 - b_1 - ka_1)\right] + \\
& \quad + \bar{I}_{FH}\left[\frac{B^{(2)}}{4}\left(a_0 b_1 - a_1\left(\theta_0 + \varphi^{(2)} - ka_0\right)\right) + \right. \\
& \qquad\left. + \mu\frac{B^{(1)}}{8}\left(3a_1\left(\theta_2 - \frac{5}{3}a_1 - kb_1\right) - b_1(\theta_1 - b_1 - ka_1)\right) - a_1\lambda_0^{(1)}\right] - \\
& \quad - \mu(\theta_0 - ka_0)\frac{\lambda_0^{(1)}}{2} - a_0\frac{\lambda_{a1}^{(2)}}{4} - \mu a_1\lambda_{b1}^{(1)} - (\theta_0 - ka_0)\frac{\lambda_{b1}^{(2)}}{4} - \\
& \quad - \frac{\lambda_{b1}^{tw(2)}}{4} - \mu\frac{\lambda_0^{tw(1)}}{2} - \frac{1}{2}\Lambda_b^2 \Bigg\}
\end{aligned}
$$

(7.7)

It is assumed here that the azimuthal inhomogeneity of inflow parameters λ is limited by first harmonics; higher harmonics are ignored.

So that, the mathematical expression of the rotor longitudinal force at specified rotor conditions was found in terms of characteristic measures of non-uniform parameters.

7.3.6   It is useful for further discussion to analyze the terms $(b_1 + ka_1 - \theta_1)$ and $(a_1 - kb_1 + \theta_2)$. The found expressions (5.21) for blade cone tilt angles a_1, b_1 can be rewritten in the following way with extracted terms related to θ_1, θ_2:

$$a_1 = D_0 A_{10} + D_{k2} B_{10} - D_0 \theta_2 \left(\frac{B_1^{(4)}}{B_2^{(4)}} + 3\frac{\epsilon\mu\mu^2}{2} \right) + D_{k2}\theta_1 \left(\frac{B_1^{(4)}}{B_2^{(4)}} + \frac{\epsilon\mu\mu^2}{2} \right),$$

$$b_1 = D_{0\mu} B_{10} - D_k A_{01} + D_{0\mu}\theta_1 \left(\frac{B_1^{(4)}}{B_2^{(4)}} + \frac{\epsilon\mu\mu^2}{2} \right) + D_k \theta_2 \left(\frac{B_1^{(4)}}{B_2^{(4)}} + 3\frac{\epsilon\mu\mu^2}{2} \right),$$

where

$$A_{10} := \mu \frac{8B_1^{(3)}}{3B_2^{(4)}} \left(\theta_0 - ka_0 + \varphi_1^{(3)} \right) +$$

$$+ \frac{4}{B_2^{(4)}} \left(\frac{\lambda_{b1}^{(3,1)}}{3} + \mu\frac{\lambda_0^{(2,1)}}{2} + \frac{B_1^{(4)}}{4}\bar{\Omega}_x - \frac{2\bar{\Omega}_y - \bar{\epsilon}_{\Omega x}}{\gamma} \right),$$

$$B_{10} := \mu \frac{4B_1^{(3)}}{3B_2^{(4)}} a_0 - \frac{4}{B_2^{(4)}} \left(\frac{\lambda_{a1}^{(3,1)}}{3} + \frac{B_1^{(4)}}{4}\bar{\Omega}_y + \frac{2\bar{\Omega}_x + \bar{\epsilon}_{\Omega y}}{\gamma} \right).$$

With such interpretation, these terms can be expressed as:

$$b_1 + ka_1 - \theta_1 = \left(D_{0\mu} + kD_{k2} \right) B_{10} - \bar{l}_{FH} D_0 D_k^{l_{FH}} A_{10} +$$

$$+ \theta_1 \bar{l}_{FH} D_0 \left\{ \frac{4B_1^{(3)}}{3B_2^{(4)}} \left(1 + k^2 - (1 - k^2)\frac{\epsilon\mu\mu^2}{1 + \frac{\epsilon\mu\mu^2}{2}} \right) - \right.$$

$$\left. - D_k^{l_{FH}} \left(k\left(1 + 3\frac{\epsilon\mu\mu^2}{2} \right) + \bar{l}_{FH}\frac{4\epsilon_I}{B_2^{(4)}\gamma} \right) \right\} + \tag{7.8a}$$

$$+ \theta_2 \bar{l}_{FH} D_0 D_k^{l_{FH}} \left[1 + 3\frac{\epsilon\mu\mu^2}{2} + \bar{l}_{FH}\frac{4B_1^{(3)}}{3B_2^{(4)}} \right],$$

$$a_1 - kb_1 + \theta_2 = (D_0 + kD_k)A_{10} + D_0 \left(\frac{2k\epsilon_\mu \mu^2}{1 + \frac{\epsilon_\mu \mu^2}{2}} + \bar{l}_{FH} D_k^{l_{FH}} \right) B_{10} -$$

$$- 2\epsilon_\mu \mu^2 D_0 (\theta_2 - k\theta_1) - \theta_2 \bar{l}_{FH} D_0 \left[\frac{4B_1^{(3)}}{3B_2^{(4)}} - (k + \bar{l}_{FH} D_k^{l_{FH}}) \frac{4\epsilon_I}{B_2^{(4)} \gamma} \right] +$$

$$+ \theta_1 \bar{l}_{FH} D_0 \left[\frac{4\epsilon_I}{B_2^{(4)} \gamma} \left(1 + \frac{\bar{l}_{FH} \frac{4B_1^{(3)}}{3B_2^{(4)}}}{1 + \frac{\epsilon_\mu \mu^2}{2}} \right) + k \frac{4B_1^{(3)}}{3B_2^{(4)}} \left(1 + \frac{2\epsilon_\mu \mu^2 + \bar{l}_{FH} \frac{4B_1^{(3)}}{3B_2^{(4)}}}{1 + \frac{\epsilon_\mu \mu^2}{2}} \right) \right].$$

$$(7.8b)$$

These terms equal zero at hovering or vertical motion ($\mu = 0$) without rotation of the rotor system ($\Omega_x = 0$, $\Omega_y = 0$) and with negligibly small flapping hinge offset ($l_{FH} = 0$); in this case $A_{10} = 0$ and $B_{10} = 0$. The term B_{10} is dominant for $b_1 + ka_1 - \theta_1$ at forward motion ($\mu > 0$), the term A_{10} is dominant for $a_1 - kb_1 + \theta_2$. The terms show dependence on cyclic pitch with longer flapping hinge offset ($l_{FH} > 0$). The term $a_1 - kb_1 + \theta_2$ depends on cyclic pitch stronger at forward motion, that is indicated by $2\epsilon_\mu \mu^2 D_0 (\theta_2 - k\theta_1)$.

7.3.7   The longitudinal rotor force (7.7) is caused only by longitudinal tilt of the total lift force due to the longitudinal tilt a_1 of the blade cone at hovering or vertical motion ($\mu = 0$) without rotor system rotation ($\Omega_x = 0$, $\Omega_y = 0$) and with negligibly small flapping hinge offset:

$$C_H = a_1 C_T,$$

the terms $(b_1 + ka_1 - \theta_1)$ and $(a_1 - kb_1 + \theta_2)$ equal zero at such conditions as shown above (7.8). Here is taken into account that the inflow ratio is azimuthally homogenous at hovering or vertical motion and harmonics of λ parameters equal zero.

The profile drag forces of the rotor blades contribute to the longitudinal rotor force at forward motion, that is represented by the second summand in (7.7). The profile drag component linearly increases with the forward motion velocity ($C_H \propto \mu$) and is directed opposite the forward motion. The impact of the azimuthally inhomogeneous induced drag increases in forward motion. The induced airflow becomes inhomogeneous at forward motion and contributes to the longitudinal rotor force as well.

The longitudinal force responds to longitudinal rotor system rotation (Ω_y) implicitly by the longitudinal blade cone response ($a_1 \propto -\Omega_y$), as was shown in 5.4.6, and explicitly by the change of attack angles of front and back blades at such rotation. The explicit response is represented in (7.7) by the summand with the Ω_y multiplier. This explicit response is caused by difference in projections of blade lift forces on the rotor longitudinal direction; the projections are caused by the blade cone angle a_0, which provokes the tilt of the blade lift forces toward the rotor hub center. The difference in the lift forces and their projections are caused by different blade element attack angles of front and back blades due to the different circumferential velocities caused by the rotor system rotation. This explicit response is directed opposite to the rotor system rotation and proportionally depends on the blade cone angle a_0.

The longitudinal force responds to lateral rotor system rotation (Ω_x), that is represented by the component with the Ω_x multiplier in (7.7). This response is caused by the different induced drag of advancing and retreating blades due to different element attack angles; the difference in the attack angles is caused by different circumferential velocities due to the lateral rotor system rotation. The longitudinal force responds to such rotation toward direction of the vector of this angular velocity i.e. perpendicularly to the rotation.

The impact of the flapping hinge offset on the longitudinal force is generally caused by the fact, that the rotor disc part between the rotor hub and blade flapping hinges does not create any lift force and does not participate in flapping motion, that causes decrease of the lift force projection on the rotor plane. This impact is represented explicitly by the terms with the multiplier $\bar{l}_{FH}$ in (7.7) and implicitly by the terms with $(b_1 + ka_1 - \theta_1)$ and $(a_1 - kb_1 + \theta_2)$. This contribution of the hinge offset mostly depends on the blade cone tilt.

7.4   ROTOR SIDE FORCE

7.4.1   A side force of the rotor system is a component of the total rotor force, which is applied to the hub center (the point O_r), is directed along the lateral y_r-axis of the rotor frame in the rotor plane, and is denoted S (fig. 7.1). The side force is directed along $\pi/2$ azimuthal direction. The side force is much smaller than the rotor thrust; however, this force enables control over the rotorcraft in its lateral direction due to ability to be changed according to swashplate tilt and to create a lateral moment about the rotorcraft center of mass.

7.4.2   The side rotor force can be derived by integrating the expression of S (7.1) with the element aerodynamic force components (3.17). The achieved expression can be split into two parts: a part caused by blade lift forces; and a part caused by blade profile drag:

$$
S = \frac{\rho n S_B \bar{C}_L^\alpha (\omega R)^2}{2} \left[\frac{1}{2\pi} \int_0^{2\pi} \int_0^{L_B} \frac{c(y_b) C_L^\alpha(y_b)}{S_B \bar{C}_L^\alpha} (\varphi_e \bar{V}_{ex} + \bar{V}_{ez}) \times \right.
$$

$$
\left. \times (-\bar{V}_{ex} \sin\psi \sin\beta + \bar{V}_{ez} \cos\psi) \, dy_b d\psi \right] - \tag{7.9}
$$

$$
- \frac{\rho n S_B (\omega R)^2}{2} \frac{1}{2\pi} \int_0^{2\pi} \int_0^{L_B} \frac{c(y_b) C_D(y_b)}{S_B} \bar{V}_{ex} (\bar{V}_{ex} \cos\psi + \bar{V}_{ey} \sin\psi \cos\beta) \, dy_b d\psi.
$$

The first summand represents a sum of projections of element lift forces of all blades on the lateral y_r-axis of the rotor frame; the second summand represents a sum of projections of profile drag of all blades on this y_r-axis.

7.4.3   A side force coefficient C_S is introduced to generalize the side rotor force over the rotor system size, the rotor shaft angular speed and air density. The coefficient C_S is defined as the side rotor force normalized to the rotor disc area (πR^2) and to dynamic pressure of airflow on a blade tip at circumferential motion $(\rho(\omega R)^2/2)$:

$$
C_S := \frac{S}{\frac{1}{2}\rho\pi R^2(\omega R)^2} = \sigma \bar{C}_L^\alpha \left[\frac{1}{2\pi} \int_0^{2\pi} \int_0^{L_B} \frac{c(y_b)C_L^\alpha(y_b)}{S_B\bar{C}_L^\alpha} (\varphi_e\bar{V}_{ex} + \bar{V}_{ez}) \times \right.
$$

$$
\left. \times (-\bar{V}_{ex}\sin\psi\sin\beta + \bar{V}_{ez}\cos\psi)\,dy_b\,d\psi \right] - \quad (7.10)
$$

$$
- \sigma\left[\frac{1}{2\pi} \int_0^{2\pi} \int_0^{L_B} \frac{c(y_b)C_D(y_b)}{S_B} \bar{V}_{ex}(\bar{V}_{ex}\cos\psi + \bar{V}_{ey}\sin\psi\cos\beta)\,dy_b\,d\psi \right].
$$

7.4.4   In order to explain the structure of the side rotor force, the term $(-\bar{V}_{ex}\sin\psi\sin\beta + \bar{V}_{ez}\cos\psi)$ can be decomposed after expanding the normalized coordinates of element air velocity according to (3.6) with the assumption of the small blade flapping angle $(\sin\beta \approx \beta,\ \cos\beta \approx 1)$ and with limitation of blade flapping motion by first harmonics (5.12). The summands of an achieved expression can be reordered in the following way:

$$
-\bar{V}_{ex}\sin\psi\sin\beta + \bar{V}_{ez}\cos\psi = b_1\bar{V}_{ex} - a_0\bar{V}_{ex}\sin\psi -
$$

$$
- \cos\psi\left(\mu(a_0\cos\psi - a_1) - \lambda(y_b,\psi) - \frac{l_{FH} + y_b}{R}(\bar{\Omega}_x\sin\psi + \bar{\Omega}_y\cos\psi) \right) -
$$

$$
- \bar{l}_{FH}\cos\psi(b_1\cos\psi - a_1\sin\psi).
$$

Inserting this expression in the expression of C_S (7.10) and splitting in accordance with these summands:

$$
C_S = b_1\left[\frac{\sigma\bar{C}_L^\alpha}{2\pi} \int_0^{2\pi} \int_0^{L_B} \frac{c(y_b)C_L^\alpha(y_b)}{S_B\bar{C}_L^\alpha} (\varphi_e\bar{V}_{ex} + \bar{V}_{ez})\bar{V}_{ex}\,dy_b\,d\psi \right] -
$$

$$
- a_0\left[\frac{\sigma\bar{C}_L^\alpha}{2\pi} \int_0^{2\pi} \int_0^{L_B} \frac{c(y_b)C_L^\alpha(y_b)}{S_B\bar{C}_L^\alpha} (\varphi_e\bar{V}_{ex} + \bar{V}_{ez})\bar{V}_{ex}\sin\psi\,dy_b\,d\psi \right] -
$$

$$
- \left[\frac{\sigma\bar{C}_L^\alpha}{2\pi} \int_0^{2\pi} \int_0^{L_B} \frac{c(y_b)C_L^\alpha(y_b)}{S_B\bar{C}_L^\alpha} (\varphi_e\bar{V}_{ex} + \bar{V}_{ez})\cos\psi \times \right.
$$

$$
\left. \times \left(\mu(a_0\cos\psi - a_1) - \lambda(y_b,\psi) - \frac{l_{FH} + y_b}{R}(\bar{\Omega}_x\sin\psi + \bar{\Omega}_y\cos\psi) \right)dy_b\,d\psi \right] - \quad (7.11)
$$

$$
- \bar{l}_{FH}\left[\frac{\sigma\bar{C}_L^\alpha}{2\pi} \int_0^{2\pi} \int_0^{L_B} \frac{c(y_b)C_L^\alpha(y_b)}{S_B\bar{C}_L^\alpha} (\varphi_e\bar{V}_{ex} + \bar{V}_{ez})\cos\psi(b_1\cos\psi - a_1\sin\psi)\,dy_b\,d\psi \right] -
$$

$$
- \sigma\frac{1}{2\pi} \int_0^{2\pi} \int_0^{L_B} \frac{c(y_b)C_D(y_b)}{S_B} \bar{V}_{ex}(\bar{V}_{ex}\cos\psi + \bar{V}_{ey}\sin\psi\cos\beta)\,dy_b\,d\psi.
$$

The side force can be decomposed into the following components according to the representation:

- The total lift force of all blades is directed close to the axis of the blade cone. If the cone has some sideways tilt ($b_1 \neq 0$), then the total lift force is tilted together with the cone axis in the rotor lateral direction, and the total lift force has a non-zero projection on the y_r-axis of the rotor frame. The first summand represents this projection; the expression in square brackets represents the total lift force of all blades, and the multiplier b_1 represents approximated $\sin b_1$. The greater is sideways blade cone tilt b_1, the stronger is the rotor side force. The total lift force approximately equals the rotor thrust (7.2); therefore, the first summand can be rewritten as $b_1 C_T$.
- Since element lift forces are directed almost along the normal z-axis of the blade-hinge frame, the lift forces of a blade with non-zero flapping angle have a non-zero projection on the rotor plane in radial direction toward the hub center. A lift force of a blade, which forms a cone with cone angle a_0, has such a projection directed radially to the hub center for any azimuthal position. If the blade lift force is azimuthally homogeneous, then averaging of such projections of the lift forces on the rotor plane due to the blade cone equals zero; if the blade lift force in the left part of the rotor disc differs from the lift force in the right part, then projections of these forces are not compensated and force difference acts in the rotor lateral direction. The greater is the cone angle a_0 at laterally inhomogeneous lift forces, the stronger is this side force component. This effect is represented by the second summand in (7.11); the multiplier $\sin \psi$ of its integrand indicates importance of the azimuthal inhomogeneity of lift forces in the rotor lateral direction. The lateral lift force inhomogeneity appears basically at forward motion ($\mu > 0$) because of different air blowing of advancing and retreaded blades.
- A lift force of a blade element is aligned along the normal z-axis with slight tilt in opposite direction to the element air velocity. Such tilt forms a projection of the lift force on the tangential x-axis, that causes the element induced drag described in 3.5.7. If the blade blowing conditions are similar for any azimuthal direction, then the average induced drag equal zero. If blade blowing conditions are longitudinally inhomogeneous, that a front blade has different induced drag than a back blade, then this induced drag force difference acts in the sideways direction and contributes to the side rotor force. This force component is represented by the third summand, where the multiplier $\cos \psi$ of the integrand indicates importance of the longitudinal inhomogeneity of the induced drag. Generally, such longitudinal inhomogeneity appears at forward motion, when the front blade is blown from below and the back blade is blown from above by the airflow of the forward motion.
- The previous summands are determined with the assumption of negligibly small flapping hinge offset. The fourth summand represents compensation force term caused by the non-flapping part of the hinge offset.
- The fifth summand represents the averaged profile drag of all elements of all

blades projected on the rotor lateral direction. An element profile drag depends only on element air velocity and acts in opposite direction to the blade motion. Averaging of profile drag projections on the lateral y_r-axis of the rotor frame is considered from the point of view of the side rotor force. This averaging can be represented by a difference between such profile drag projections, which is created by blades in the front part of the rotor disc, and profile drag projections created by blades in the back part of the disc. Since blade element air velocities of front blades have same magnitudes but are directed opposite to air velocities of elements of back blades, then the averaged profile drag in the lateral direction is always compensated and equals zero. This can be proven that the summand equals zero by integrating with expanded normalized element air velocities according to (3.6).

7.4.5   Integrating all summands (7.6) with expanded normalized element air velocities according to (3.6), the expression of the coefficient of side rotor force can be rewritten in the following way with the characteristic measures introduced in 4:

$$C_S = b_1 C_T +$$

$$+ \sigma \bar{C}_L^\alpha \left\{ -3\mu a_0 \left(\frac{B^{(2)}}{4} (\theta_0 + \varphi^{(2)} - ka_0) + \mu \frac{B^{(1)}}{2} a_1 + \frac{\lambda_0^{(1)}}{2} \right) - \right.$$

$$- (\theta_1 - b_1 - ka_1) \left[\mu \frac{B^{(2)}}{4} a_1 + \frac{\lambda_0^{(2)}}{4} + \mu \frac{\lambda_{b1}^{(1)}}{8} \right] +$$

$$+ (\theta_2 + a_1 - kb_1) \left[a_0 \left(\frac{B^{(3)}}{6} + \mu^2 \frac{B^{(1)}}{2} \right) - \mu \frac{\lambda_{a1}^{(1)}}{8} \right] -$$

$$- \bar{\Omega}_x \left[\frac{B^{(3)}}{6} a_0 + \mu \frac{B^{(2)}}{16} (\theta_1 - b_1 - ka_1) \right] +$$

$$+ \bar{\Omega}_y \left[\frac{B^{(3)}}{6} (\theta_0 + \varphi^{(3)} - ka_0) - \mu \frac{B^{(2)}}{16} (\theta_2 - 7a_1 - kb_1) + \frac{\lambda_0^{(2)}}{2} \right] -$$

$$- \bar{l}_{FH} \left[a_1 \left(\frac{B^{(2)}}{4} a_0 + \mu \frac{B^{(1)}}{8} (\theta_1 - b_1 - ka_1) \right) + \right.$$

$$\left. + b_1 \left(\frac{B^{(2)}}{4} (\theta_0 + \varphi^{(2)} - ka_0) - \mu \frac{B^{(1)}}{8} (\theta_2 - 7a_1 - kb_1) + \lambda_0^{(1)} \right) \right] +$$

$$\left. + (\theta_0 - ka_0) \frac{\lambda_{a1}^{(2)}}{4} + \frac{\lambda_{a1}^{tw(2)}}{4} + \mu a_1 \lambda_{a1}^{(1)} - a_0 \frac{\lambda_{b1}^{(2)}}{4} + \frac{1}{2} \Lambda_a^2 \right\}$$

$$(7.12)$$

It is assumed here that the azimuthal inhomogeneity of inflow parameters λ is limited by first harmonics; higher harmonics are ignored.

So that, here was found the mathematical expression for calculation of the rotor side force at specified rotor conditions in terms of characteristic measures of distributed parameters.

7.4.6   The side rotor force is determined only by lateral tilt of the total lift force due to the lateral tilt b_1 of the blade cone at hovering or vertical motion ($\mu = 0$) without rotor system rotation ($\Omega_x = 0$, $\Omega_y = 0$) and with negligibly small flapping hinge offset:

$$C_S = b_1 C_T,$$

since $b_1 + ka_1 - \theta_1 = 0$ and $a_1 - kb_1 + \theta_2 = 0$, and the inflow is azimuthally homogenous at these conditions. Together with the longitudinal rotor force at the same conditions (7.3.7), it can be generally stated, that the longitudinal and lateral rotor forces are determined by tilt of the total blade lift force, which is directed along the tilted axis of the blade cone at hovering or vertical motion. So that, the longitudinal and lateral forces can be controlled through the blade cone tilt, which can be performed via the swashplate mechanism. It gives the possibility to perform control over the rotorcraft via the swashplate mechanism.

Such direct control over these forces is distorted in a certain way at the forward motion due to azimuthally inhomogeneous profile and induced blade drags. The azimuthal inhomogeneous of the induced airflow, which appears at forward motion, contributes to such distortion as well. It must be admitted, that the blade profile drag does not contribute to the side rotor force in comparison with the longitudinal rotor force.

The side rotor force responds to lateral rotor system rotation (Ω_x) implicitly by the sideways blade cone tilt ($b_1 \propto -\Omega_x$), as was shown in 5.4.6, and explicitly by change of element attack angles of advancing and retreated blades at such rotation. This explicit response is represented by the summand with the Ω_x multiplier in (7.12). The explicit response causes by different circumferential velocities of advancing and retreated blades due to the lateral rotor system rotation; this causes the difference of element attack angles and the difference of blade lift forces of the advancing and retreating blades; the different blade lift forces have different projections on the rotor plane toward the hub center at non-zero blade cone angle ($a_0 > 0$); the difference of such lift force projections of advancing and retreating blades contributes to the side rotor force. So that, the side rotor force responds to the lateral rotor system rotation at the presence of the cone angle; the side force is directed opposite to this rotation motion; the response is proportional to the rotation velocity Ω_x with a proportionality coefficient determined by a_0. This response is similar to the longitudinal rotor force response on the longitudinal rotor system rotation, as described in 7.3.7. Generalizing over the longitudinal rotor force response on longitudinal rotation and the side rotor force response on lateral rotation, the rotor system responds to the rotor system rotation by force opposite this rotation motion; the force is proportional to angular velocity of this rotation; a proportionality coefficient is defined by blade cone angle ($a_0 B^{(3)}/6$) at hovering and vertical motion; the proportionality coefficient becomes slightly different for the longitudinal and side rotor forces at forward motion.

The side rotor force responds to longitudinal rotation of the rotor system, that is represented by the summand with the Ω_y multiplier in (7.12). This response is caused by different induced drag of front and back blades due to different elements attack angles, which are caused by different circumferential velocities due to the longitudinal rotation. This response is directed toward the vector of the longitudinal angular velocity. This response is similar to the response of the longitudinal rotor force on lateral rotor system rotation described in 7.3.7. Generalizing over the longitudinal force response on lateral rotation and the side rotor force response on longitudinal rotation, the rotor system responds to the rotor system rotation by a force acting transversely to the rotation and directed along the vector of the angular velocity $(\vec{\Omega})$; the force is proportional to the angular speed; the proportionality coefficient is determined by

collective pitch and averaged inflow ratio $\left(\dfrac{B^{(3)}}{6}(\theta_0 + \varphi^{(3)} - ka_0) + \dfrac{\lambda_0^{(2)}}{2} \right)$ at hover-

ing and vertical motion; the proportionality coefficient becomes slightly different for the longitudinal and side rotor forces at forward motion.

Similarly to the longitudinal rotor force, the impact of the flapping hinge offset on the side rotor force is caused by no lift force creation in the area between the hub center and blade flapping hinges, as well as by no participation of this area in the flapping motion of the blades. This impact is represented by the terms with the multiplier $\bar{l}_{FH}$ in (7.12) and implicitly by the terms with $(b_1 + ka_1 - \theta_1)$ and $(a_1 - kb_1 + \theta_2)$.

8 Rotor Moment on the Hub

Blades of the rotor system act on their flapping hinges by forces described in chapter 6.2. Each blade affects its flapping hinge in contacts of blade lugs with the hinge axle, on which the lugs are put. These blade forces are transferred on the rotor hub through the hinge, which has certain offset from the hub center. The impact of blades on the rotor hub is analyzed here about the center of the rotor hub, which was chosen as the reference point (see 2.4.14). Based on this approach, the blade forces on the flapping hinges create moments about the hub center with arms of the flapping hinge offsets. These moments about the hub center, which are created by blade forces on the flapping hinges, are analyzed in this chapter.

8.1 GENERAL STATEMENTS ABOUT ROTOR MOMENT

8.1.1 Forces of a blade, which act on the blade flapping hinge in all contacts of blade lugs with the hinge axle, create force moments about the hub center due to offset of the hinge axle from the hub center (the flapping hinge offset). A total moment, with which the blade acts on the hub, is useful to analyze concerning the flapping hinge center as a reference point of interaction of the blade with its hinge. The flapping hinge center is located at the intersection of the blade longitudinal axis with the axis of the flapping hinge and coincides with the origin of the blade-hinge frame, as was defined in 2.2.5.

8.1.2 Consider a blade with two lugs, which are put on a flapping hinge axle and which are denoted $lug1$ and $lug2$. The blade with its flapping hinge locates in certain azimuthal position ψ at a certain moment of time t. The blade affects the flapping hinge by forces applied to the lug contacts with the hinge axle: the first lug acts by a force, which is described by a vector $\vec{F}_{lug1}^{(r)}$ in the rotor frame, in a contact point described by a radius-vector $\vec{r}_{lag1}^{(r)}$ in the rotor frame; the second lug acts by a force $\vec{F}_{lug2}^{(r)}$ in a contact point $\vec{r}_{lag2}^{(r)}$ in the rotor frame. The blade creates an instant blade moment on the hub about the hub center, which equals the vector sum of moments created by all blade lugs:

$$\vec{M}_{blade}^{(r)} = \vec{r}_{lag1}^{(r)} \times \vec{F}_{lug1}^{(r)} + \vec{r}_{lag2}^{(r)} \times \vec{F}_{lug2}^{(r)}.$$

A radius-vector $\vec{l}_{FH}^{(r)}(-l_{FH}\cos\psi, l_{FH}\sin\psi, 0)$ represents position of the flapping hinge center in the rotor frame according to the blade azimuthal position and represents the flapping hinge offset. The blade moment can be represented in terms of the flapping hinge center as a reference point of interaction of the blade with its hinge:

$$\vec{M}_{blade}^{(r)} = \vec{l}_{FH}^{(r)} \times \left(\vec{F}_{lug1}^{(r)} + \vec{F}_{lug2}^{(r)} \right) + \left[(\vec{r}_{lag1}^{(r)} - \vec{l}_{FH}^{(r)}) \times \vec{F}_{lug1}^{(r)} + (\vec{r}_{lag2}^{(r)} - \vec{l}_{FH}^{(r)}) \times \vec{F}_{lug2}^{(r)} \right],$$

DOI: 10.1201/9781003296232-8

where $\vec{F}_{lug1}^{(r)} + \vec{F}_{lug2}^{(r)}$ represents the total blade force on the hinge $\vec{F}_{FH}^{(r)}$ with coordinates in the rotor frame, which was determined in 6.2. The moment component in the square brackets represents the blade moment on the hinge M_{onFH} defined in 2.4.11 and 5.6.3; however, this blade moment on the hinge is specified in the rotor frame here:

$$(\vec{r}_{lag1}^{(r)} - \vec{l}_{FH}^{(r)}) \times \vec{F}_{lug1}^{(r)} + (\vec{r}_{lag2}^{(r)} - \vec{l}_{FH}^{(r)}) \times \vec{F}_{lug2}^{(r)} = \vec{M}_{onFH}^{(r)},$$

where $\vec{M}_{onFH}^{(r)} := M^{(h \rightarrow r)} \vec{M}_{onFH}$ is the blade moment on the hinge with coordinates in the rotor frame. The matrix $M^{(h \rightarrow r)}$ is a transformation matrix from the blade-hinge frame to the rotor frame and can be found by transposition of the matrix from the rotor frame to the blade-hinge frame $M^{(r \rightarrow h)}$ defined in 2.2.5 (2.1): $M^{(h \rightarrow r)} = M^{(r \rightarrow h)T}$. It must be admitted that the blade moment on the flapping hinge ($\vec{M}_{onFH}$) is directed only around the normal z-axis of the blade-hinge frame; the blade does not create any moment around the y-longitudinal and x-tangential axes of the blade-hinge frame due to no friction in the flapping and feathering hinges. Therefore, the blade moment on the flapping hinge in the rotor frame can be expressed as:

$$\vec{M}_{onFH}^{(r)} = M_{onFH} \left(M^{(h \rightarrow r)} \vec{e}_z \right) \tag{8.1}$$

According to this approach, the blade affects its flapping hinge by the blade force on the hinge, which is applied to the flapping hinge center, and by the blade moment on the hinge caused by separate locations of the blade lugs. The blade creates the moment on the rotor hub about the hub center, which consists of: the moment produced by the blade force on the hinge $\vec{F}_{FH}^{(r)}$ with an arm of the hinge offset $\vec{l}_{FH}^{(r)}$, and the blade moment on the hinge $\vec{M}_{onFH}^{(r)}$:

$$\vec{M}_{blade}^{(r)} = \vec{l}_{FH}^{(r)} \times \vec{F}_{FH}^{(r)} + \vec{M}_{onFH}^{(r)}.$$

The moment $\vec{M}_{blade}^{(r)}$, with which the blade acts on the hub, is called the blade moment on the hub hereafter.

8.1.3   A rotor system has more than one flapping hinge ($n > 1$). The flapping hinges suited on the hub in the axisymmetric order. Each blade is attached to a correspondent flapping hinge. Each blade affects its flapping hinge and produces a moment on the hub due to the flapping hinge offset.

Consider a i-th blade out of n blades of the rotor system; where i is an index of the blade. The i-th blade locates in certain azimuthal position $\psi_i(t)$ at certain moment of time t. Position of the flapping hinge center of the i-th blade is described by a radius-vector $\vec{l}_{FH}^{(r)}(\psi_i(t))$ with coordinates in the rotor frame according to the azimuthal position of the blade. This blade applies the force on its hinge $\vec{F}_{FHi}^{(r)}(t)$ in the hinge center and the moment on its hinge $\vec{M}_{onFHi}^{(r)}(t)$ with coordinates in the rotor frame. So, the i-th blade creates a moment on the hub:

$$\vec{M}_{blade(i)}^{(r)} = \vec{l}_{FH}^{(r)}(\psi_i(t)) \times \vec{F}_{FHi}^{(r)}(t) + \vec{M}_{onFHi}^{(r)}(t).$$

The blade moment on the hub periodically repeats every full turn of the rotor shaft at steady rotation with $Period = 2\pi/\omega$; the moment of any blade is same in same blade azimuthal position at the steady rotation. Based on this, every blade repeats the moment of its previous blade passing same azimuthal position with the time delay $Period/n$. Therefore, a hub moment of any i-th blade in position $\psi_i(t)$ can be described by a moment of one selected blade as a reference (let it is $i = 1$) in the same azimuthal position $\psi_i(t) = \psi_1(t + (i-1)Period/n)$ with taking into account the time delay $(i-1)Period/n$:

$$\vec{M}^{(r)}_{blade(i)} = \vec{l}^{(r)}_{FH}\left(\psi_1\left(t + \frac{(i-1)Period}{n}\right)\right) \times \vec{F}^{(r)}_{FH1}\left(t + \frac{(i-1)Period}{n}\right) + $$
$$+ \vec{M}^{(r)}_{onFH1}\left(t + \frac{(i-1)Period}{n}\right).$$

8.1.4   The instant total moment on the hub equals to a superposition of hub moments of all blades at certain moment of time t:

$$\vec{M}^{(r)}_{HUBinstant} = \sum_{i=1}^{n} \vec{M}^{(r)}_{blade(i)} = \sum_{i=1}^{n}\left(\vec{l}^{(r)}_{FH}(\psi_i(t)) \times \vec{F}^{(r)}_{FHi}(t) + \vec{M}^{(r)}_{onFHi}(t)\right).$$

If the blades act on their hinges by equal forces and create equal moments on the hinges, then the total instant moment on the hub equals zero due to the axisymmetric order of the hinges. If forces and moments on the hinges of the rotor blades are different, then the total instant moment is non-zero and affects the rotor hub. These blade forces and moments are different in different azimuthal positions if the blade force and the blade moment on the blade hinge depend on blade azimuthal position. The instant total moment on the hub repeats at steady blade rotation with a period, which is the number of blades (n) times less than the period of one full rotor turn: $Period/n$. This creates vibrations on the hub, which transfers to the rotorcraft.

8.1.5   Analogously to the total rotor force, the total rotor hub moment is defined as the averaged instant total moment on the hub over one full shaft turn. The total rotor hub moment can be determined by integration of the instant total moment on the hub over one period of the shaft turn divided on the shaft turn period:

$$\vec{M}_{HUB} := \frac{\int\limits_{t}^{t+Period} \vec{M}_{HUBinstant}(t)dt}{Period} =$$
$$= \frac{1}{Period}\sum_{i=1}^{n}\int\limits_{t}^{t+Period}\left(\vec{l}^{(r)}_{FH}(\psi_i(t)) \times \vec{F}^{(r)}_{FHi}(t) + \vec{M}^{(r)}_{onFHi}(t)\right)dt.$$

where $\vec{M}_{HUB}$ is the total hub moment with coordinates in the rotor frame. This approach ignores the vibration created by the blade moments on the hub. As it was shown above, that one blade can describe any other blade at same azimuthal position and with the correspondent time delay, then the total instant moment can be determined by only one reference blade (let it has index 1) with the correspondent time delays:

$$\vec{M}_{HUB} = \frac{1}{Period} \sum_{i=1}^{n} \int_{t}^{t+Period} \left[\vec{l}_{FH}^{(r)} \left(\psi_1 \left(t + \frac{i-1}{n} Period \right) \right) \times \vec{F}_{FH1}^{(r)} \left(t + \frac{i-1}{n} Period \right) + \right.$$

$$\left. + \vec{M}_{onFH1}^{(r)} \left(t + \frac{i-1}{n} Period \right) \right] dt.$$

Since the force $\vec{F}_{FH1}^{(r)}$ and the moment $\vec{M}_{onFH1}^{(r)}$ on the hinge of the blade ($i = 1$) are periodic and are changed synchronically with the blade azimuthal position at steady rotation, then the integral over one period of blade rotation does not depend on time and the blade time delay $Period(i-1)/n$. This integral does not depend on a blade index at steady rotation and is the same for any blade. As soon as such integrals are the same for all blades, then just an integral for one blade (for instance $i = 1$) can be used:

$$\vec{M}_{HUB} = \frac{n}{Period} \int_{t}^{t+Period} \left(\vec{l}_{FH}^{(r)} (\psi_1(t)) \times \vec{F}_{FH1}^{(r)}(t) + \vec{M}_{onFH1}^{(r)}(t) \right) dt,$$

further the index of the blade is ignored.

Essentially, components of the blade force and moment on the hinge are determined according to blade azimuthal position: $\vec{F}_{FH}^{(r)}(\psi)$, $\vec{M}_{onFH1}^{(r)}(\psi)$ (see 6.2 and 5.6.3). The integral is convenient to present in terms of a variable of blade azimuthal position ψ instead of the variable of time t based on the relationship $\psi = \omega t$. In this case, the boundaries of integration are limited by the full round of azimuthal positions starting from convenient value of 0 and ending by 2π:

$$\vec{M}_{HUB} = \frac{n}{2\pi} \int_{0}^{2\pi} \left(\vec{l}_{FH}^{(r)}(\psi) \times \vec{F}_{FH}^{(r)}(\psi) + \vec{M}_{onFH}^{(r)}(\psi) \right) d\psi.$$

8.1.6   The coordinates of the total hub moment $\vec{M}_{HUB}(M_{HUBx}, M_{HUBy}, M_{HUBz})$ in the rotor frame, which are shown in fig. 8.1, are found on the base of the coordinates of the blade force on the hinge in the rotor frame $\vec{F}_{FH}^{(r)}(F_{FHx}^{(r)}, F_{FHy}^{(r)}, F_{FHz}^{(r)})$ according to (6.3), the coordinates of the radius-vector the flapping hinge offset $\vec{l}_{FH}^{(r)}(\psi)(-l_{FH}\cos\psi, l_{FH}\sin\psi, 0)$ at the blade azimuthal position ψ in the rotor frame as the arm of the blade force, and coordinates of the blade moment on the hinge in the rotor frame $\vec{M}_{onFH}^{(r)}$, as was shown in (8.1):

$$M_{HUBx} = \frac{n}{2\pi} \int_{0}^{2\pi} \left[\vec{l}_{FH}^{(r)}(\psi) \times \vec{F}_{FH}^{(r)}(\psi) + \vec{M}_{onFH}^{(r)}(\psi) \right]_{x} d\psi =$$

$$\text{(8.2a)}$$

$$= \frac{n}{2\pi} \int_{0}^{2\pi} \left(l_{FH} F_{FHz}^{(r)}(\psi) \sin\psi - M_{onFH} \sin\beta \cos\psi \right) d\psi,$$

$$M_{HUBy} = \frac{n}{2\pi} \int_0^{2\pi} \left[\vec{l}_{FH}^{(r)}(\psi) \times \vec{F}_{FH}^{(r)}(\psi) + \vec{M}_{onFH}^{(r)}(\psi) \right]_y d\psi =$$

$$= \frac{n}{2\pi} \int_0^{2\pi} \left(l_{FH} F_{FHz}^{(r)}(\psi) \cos\psi + M_{onFH} \sin\beta \sin\psi \right) d\psi, \qquad (8.2b)$$

$$M_{HUBz} = \frac{n}{2\pi} \int_0^{2\pi} \left[\vec{l}_{FH}^{(r)}(\psi) \times \vec{F}_{FH}^{(r)}(\psi) + \vec{M}_{onFH}^{(r)}(\psi) \right]_z d\psi =$$

$$= \frac{n}{2\pi} \int_0^{2\pi} \left(-l_{FH} \left(F_{FHy}^{(r)} \cos\psi + F_{FHx}^{(r)} \sin\psi \right) + M_{onFH} \cos\beta \right) d\psi. \qquad (8.2c)$$

The coordinate M_{HUBx} represents a lateral hub moment and acts around the longitudinal x_r-axis of the rotor frame; the coordinate M_{HUBy} represents a longitudinal hub moment and acts around the lateral y_r-axis of the rotor frame; the coordinate M_{HUBz} generally represents a resistance moment to the rotor shaft rotation (at no autorotation) and acts around the rotor axis coinciding with the z_r-axis of the rotor frame. The integrand components $M_{onFH} \sin\beta \cos\psi$ in M_{HUBx} (8.2a) and $M_{onFH} \sin\beta \sin\psi$ in M_{HUBy} (8.2b) are negligibly small in comparison to the correspondent summands due to small flapping angle β and are neglected in further analysis. With this assumption, the lateral and longitudinal hub moments are straightly determined by the flapping hinge offset: the longer is the hinge offset, the stronger are the lateral and longitudinal hub moments.

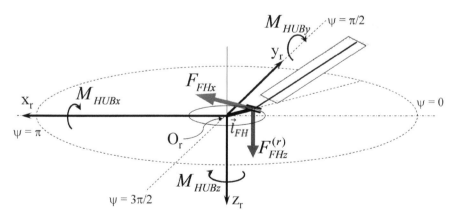

Figure 8.1   Scheme of components of rotor moment on the hub in rotor frame.

8.2   LATERAL AND LONGITUDINAL HUB MOMENTS

8.2.1   Lateral M_{HUBx} and longitudinal M_{HUBy} hub moments participate in the rotorcraft control in longitudinal and lateral directions jointly with the moments created by the rotor longitudinal force H and the rotor side force S about the rotorcraft center of mass. The lateral and longitudinal moments are determined by the vertical component $F_{FHz}^{(r)}$ of the blade forces on the flapping hinges according to (8.2a) and (8.2b), specifically, by the azimuthal distribution of this force component. Azimuthal inhomogeneity of the vertical component of a blade force on a hinge, which means dependence of the force component on blade azimuthal position, causes the lateral and longitudinal moments. If the vertical force component is longitudinally homogeneous, then moments created by all blades on the hub around the y_r-axis are compensated, and the longitudinal hub moment equals zero. If this force component is laterally homogeneous, then moments created by all blades on the hub around the x_r-axis are compensated, and the lateral hub moment equals zero.

8.2.2   The vertical component of a blade force on a flapping hinge $F_{FHz}^{(r)}$ in (8.2) is considered as a dominant reason for the lateral and longitudinal hub moments; since the contribution of a blade moment on a hinge M_{onFH} into these moments is negligibly small due to small flapping angle. This vertical component can be expressed with expanded inertial forces, which act on the blade along z_r-axis of the rotor frame according to (6.3):

$$
\begin{aligned}
F_{FHz}^{(r)} = &\int_0^{L_B} (dZ\cos\beta - dY\sin\beta) + \\
&+ J\cos\beta \frac{d^2\beta}{dt^2} - (J\cos\beta + m_B l_{FH})\left(\frac{d\Omega_x}{dt}\sin\psi + \frac{d\Omega_y}{dt}\cos\psi\right) - \\
&- 2\omega(J\cos\beta + m_B l_{FH})(\Omega_x\cos\psi - \Omega_y\sin\psi) + \\
&+ 2J\sin\beta \frac{d\beta}{dt}(\Omega_x\sin\psi + \Omega_y\cos\psi) - J\sin\beta(\Omega_x^2 + \Omega_y^2) - J\sin\beta\left(\frac{d\beta}{dt}\right)^2.
\end{aligned}
\tag{8.3}
$$

The first summand represents a total aerodynamic force of the blade acting on its flapping hinge along the z_r-axis of the rotor frame, which consists of a projection of the total blade lift force on the rotor axis $(dZ\cos\beta)$ and a projection of the total blade profile drag acting along the blade longitudinal axis at air blowing along the blade longitudinal axis $(-dY\sin\beta)$. This first summand in (8.3) can be expressed with the extended element aerodynamic force components with normalized element air velocity coordinates according to (3.17):

$$
\begin{aligned}
\int_0^{L_B} (dZ\cos\beta - dY\sin\beta) = &-\frac{\rho(\omega R)^2}{2}\int_0^{L_B}\Bigg(c(y_b)C_L^\alpha(y_b)\cos\beta\left(\varphi_e\bar{V}_{ex}^2 + \bar{V}_{ex}\bar{V}_{ez}\right) - \\
&- c(y_b)C_D(y_b)\bar{V}_{ex}\bar{V}_{ey}\sin\beta\Bigg)dy_b.
\end{aligned}
$$

The profile drag coefficient is two orders of magnitude less than the lift coefficient slope: $C_D(y_b) \ll C_L^\alpha(y_b)$. Based on this and on the assumption of small blade flapping angle $\sin\beta \approx \beta$ and $\cos\beta \approx 1$, that causes $\sin\beta \ll \cos\beta$, the profile drag term (the second term) of the integrand is much smaller than the lift force term (the first term) and can be neglected. With these, the first summand of (8.3) can be written in a simplified form:

$$\int_0^{L_B} (dZ\cos\beta - dY\sin\beta) \approx -\frac{\rho(\omega R)^2 S_B \bar{C}_L^\alpha}{2} \int_0^{L_B} \frac{c(y_b)C_L^\alpha(y_b)}{S_B \bar{C}_L^\alpha} \left(\varphi_e \bar{V}_{ex}^2 + \bar{V}_{ex}\bar{V}_{ez}\right) dy_b.$$

The second summand of the vertical component of the force on the flapping hinge (8.3) represents inertial forces due to accelerated flapping motion. The third summand represents inertial forces due to angular acceleration of the rotor system. The fourth summand represents Coriolis forces due to mutual rotation of the rotor system with the blade rotation around the rotor shaft. The fifth summand represents a projection on the rotor axis at angle $\pi/2 - \beta$ of Coriolis forces due to mutual rotation of the rotor system with flapping motion; this term can be neglected due to small flapping angle as well as due to small magnitude of angular velocities of the rotor system rotation ($|\Omega_x| \ll |\omega|$, $|\Omega_x| \ll |\omega|$). The six summand represents a projection of centrifugal forces due to rotor system rotation; this summand can be neglected due to small flapping angle as well as due to small magnitude of angular velocities of the rotor system ($|\Omega_x| \ll |\omega|$, $|\Omega_x| \ll |\omega|$). The seventh summand represents a projection on the rotor axis at angle $\pi/2 - \beta$ of centrifugal forces due to flapping motion around the flapping hinge; this summand can be neglected due to small flapping angle.

The expression for the vertical component of the force on the flapping hinge can be rewritten in a simplified form with all of the specified above assumptions and with $\cos\beta \approx 1$:

$$F_{FHz}^{(r)} = -\frac{\rho(\omega R)^2 S_B \bar{C}_L^\alpha}{2} \int_0^{L_B} \frac{c(y_b)C_L^\alpha(y_b)}{S_B \bar{C}_L^\alpha} \left(\varphi_e \bar{V}_{ex}^2 + \bar{V}_{ex}\bar{V}_{ez}\right) dy_b + J\frac{d^2\beta}{dt^2} -$$

$$- \omega^2(J + m_B l_{FH})\left((2\bar{\Omega}_x + \bar{\varepsilon}_{\Omega y})\cos\psi - (2\bar{\Omega}_y - \bar{\varepsilon}_{\Omega x})\sin\psi\right). \tag{8.4}$$

8.2.3   The lateral and longitudinal hub moments (8.2) are analyzed with the found appropriate expression for the vertical force on the flapping hinge (8.4) in terms of impact of inertial forces assuming that flapping motion is limited by first harmonics (5.12):

$$M_{HUBx} = l_{FH}\frac{nJ\omega^2}{2}\left[b_1 + \left(1 + \frac{m_B}{J}l_{FH}\right)(2\bar{\Omega}_y - \bar{\varepsilon}_{\Omega x})\right] -$$

$$- l_{FH}\frac{\rho(\omega R)^2 S_B \bar{C}_L^\alpha}{2}\frac{n}{2\pi}\int_0^{2\pi}\int_0^{L_B} \frac{c(y_b)C_L^\alpha(y_b)}{S_B \bar{C}_L^\alpha}\left(\varphi_e \bar{V}_{ex}^2 + \bar{V}_{ex}\bar{V}_{ez}\right)\sin\psi\, dy_b d\psi, \tag{8.5a}$$

$$M_{HUBy} = l_{FH}\frac{nJ\omega^2}{2}\left[a_1 - \left(1 + \frac{m_B}{J}l_{FH}\right)\left(2\bar{\Omega}_x + \bar{\varepsilon}_{\Omega y}\right)\right] -$$

$$- l_{FH}\frac{\rho(\omega R)^2 S_B \bar{C}_L^\alpha}{2}\frac{n}{2\pi}\int_0^{2\pi}\int_0^{L_B}\frac{c(y_b)C_L^\alpha(y_b)}{S_B\bar{C}_L^\alpha}\left(\varphi_e\bar{V}_{ex}^2 + \bar{V}_{ex}\bar{V}_{ez}\right)\cos\psi\,dy_b d\psi. \tag{8.5b}$$

An important aspect of the lateral and longitudinal hub moments is the dominant contribution to these moments of blade inertial forces, which are represented by the summands with square brackets in (8.5). The hub moments due to the blade inertial forces depend on the mass properties of a blade (blade mass m_B and blade mass moment J), on the number of blades in the rotor system n, on the angular speed of the rotation around the rotor shaft (ω^2), and on the flapping hinge offset (l_{FH}). This contribution of blade inertial forces in the hub moments can be split into a component due to accelerated blade flapping motion with cone tilt angles (a_1 and b_1), a component due to Coriolis forces of rotor system rotation ($\bar{\Omega}_x$, $\bar{\Omega}_y$) accompanied by the shaft rotation, and a component due to accelerated rotation of the rotor system ($\bar{\varepsilon}_{\Omega x}$, $\bar{\varepsilon}_{\Omega y}$).

The accelerated flapping motion of blades creates an azimuthally inhomogeneous force on flapping hinges in case of tilt of a blade cone. If a blade cone is not tilted and the cone axis coincides with the rotor axis, then there is no accelerated flapping motion because of constant blade flapping angle ($\beta = const$); therefore, there are no inertial forces due to accelerated flapping to create a force on the hinge; and no hub moment is created. If there is some blade cone tilt, then: (i) a force on a hinge due to the accelerated flapping is directed downward at azimuthal position with minimal flapping angle, because there is highest flapping acceleration upward; (ii) the force on the hinge is directed upward at azimuthal position with maximal flapping angle due to highest flapping acceleration downward; (iii) therefore, the force on the hinge is azimuthally inhomogeneous and creates a moment directed toward the cone tilt. So that, the blade inertial force moments encourage the rotor axis to coincide with the cone axis. The hub lateral moment appears at sideways tilt of the blade cone ($M_{HUBx} \propto b_1$); the longitudinal hub moment appears at longitudinal cone tilt ($M_{HUBy} \propto a_1$).

Rotation of the rotor system accompanied with the rotation around the rotor shaft causes Coriolis forces (described in 3.6.6) which act on each blade element along the normal z-axis of the blade-hinge frame, that is close to the vertical direction (the z_r-axis) due to small blade flapping angle. These Coriolis forces are transferred on a blade flapping hinge and contribute to hub moments. The hub moment component due to the Coriolis forces creates a transverse moment with respect to the rotor system rotation: if the rotor system rotates around the rotor longitudinal x_r-axis, then this moment acts around the rotor lateral y_r-axis ($M_{HUBy} \propto -\Omega_x$); if the rotor system rotates around the rotor lateral y_r-axis, then this moment acts around the rotor longitudinal x_r-axis ($M_{HUBx} \propto \Omega_y$). It can be interpreted that the moment due to the Coriolis forces act in azimuthal direction with $\pi/2$ ahead from the azimuthal direction of the rotor system rotation.

Accelerated rotation of the rotor system causes the inertial forces of a blade, which transfers on its flapping hinge and contributes to the hub moment. The hub moment component due to accelerated rotation acts in the opposite direction to the angular acceleration ($\varepsilon_{\Omega x}$, $\varepsilon_{\Omega y}$) and damps the change of the rotor system angular velocity.

8.2.4   Final expressions for the longitudinal and lateral hub moments are achieved by integrating aerodynamic force summands (8.5) with expanded normalized element air velocities according to (3.6) and with the characteristic measures introduced in 4:

$$M_{HUBx} = l_{FH}\frac{nJ\omega^2}{2}\left[b_1 + \left(1 + \frac{m_B}{J}l_{FH}\right)\left(2\bar{\Omega}_y - \bar{\varepsilon}_{\Omega x}\right)\right] +$$

$$+ l_{FH}\frac{\rho(\omega R)^2 nS_B\bar{C}_L^\alpha}{2}\left\{\left(\frac{B^{(3)}}{6} + 3\mu^2\frac{B^{(1)}}{8}\right)(\theta_2 + a_1 - kb_1) - \right.$$

$$\left. - \mu\left(\frac{B^{(2)}}{2}(\theta_0 + \varphi^{(2)} - ka_0) + \frac{\lambda_0^{(1)}}{2}\right) - \mu^2\frac{B^{(1)}}{2}a_1 - \frac{B^{(2)}}{4}\bar{l}_{FH}a_1 - \frac{B^{(3)}}{6}\bar{\Omega}_x - \frac{\lambda_{b1}^{(2)}}{4}\right\},$$

$$M_{HUBy} = l_{FH}\frac{nJ\omega^2}{2}\left[a_1 - \left(1 + \frac{m_B}{J}l_{FH}\right)\left(2\bar{\Omega}_x + \bar{\varepsilon}_{\Omega y}\right)\right] +$$

$$+ l_{FH}\frac{\rho(\omega R)^2 nS_B\bar{C}_L^\alpha}{2}\left\{\left(\frac{B^{(3)}}{6} + \mu^2\frac{B^{(1)}}{8}\right)(\theta_1 - b_1 - ka_1) + \right.$$

$$\left. + \mu\frac{B^{(2)}}{4}a_0 + \frac{B^{(2)}}{4}\bar{l}_{FH}b_1 - \frac{B^{(3)}}{6}\bar{\Omega}_y - \frac{\lambda_{a1}^{(2)}}{4}\right\}.$$

$$(8.6)$$

The impact of the aerodynamic forces on the longitudinal and lateral hub moments basically appears at forward motion, when azimuthal inhomogeneity of the blade lift force becomes stronger and influences these hub moments. Azimuthal inhomogeneity of the induced airflow affects as well these hub moments at forward motion, when the induced airflow inhomogeneity becomes stronger, that is indicated by terms $\lambda_{a1}^{(2)}$ and $\lambda_{b1}^{(2)}$. Generally, the impact of the inertial forces dominates in comparison to the aerodynamic force contribution.

8.2.5   Coefficients of lateral and longitudinal hub moments are introduced in order to generalize the lateral and longitudinal moments over the rotor system size, the rotor shaft angular speed, and air density. The lateral C_{mx} and longitudinal C_{my} hub moment coefficients are defined as the correspondent lateral M_{HUBx} and longitudinal M_{HUBy} hub moments normalized to the area of the rotor disc (πR^2), to dynamic pressure of airflow on a blade tip at circumferential motion ($\rho(\omega R)^2/2$), and to the radius of the rotor disc (R) as rotor system reference length:

$$C_{mx} := \frac{M_{HUBx}}{\frac{1}{2}\rho\pi R^3(\omega R)^2} = \bar{l}_{FH}\sigma\bar{C}_L^{\alpha}\left\{\frac{JR}{2\gamma I_{zz}^{(r)}}\left[b_1 + \left(1+\frac{m_B}{J}l_{FH}\right)\left(2\bar{\Omega}_y - \bar{\varepsilon}_{\Omega x}\right)\right] + \right.$$

$$+ \left(\frac{B^{(3)}}{6} + 3\mu^2\frac{B^{(1)}}{8}\right)(\theta_2 + a_1 - kb_1) -$$

$$- \mu\left(\frac{B^{(2)}}{2}(\theta_0 + \varphi^{(2)} - ka_0) + \frac{\lambda_0^{(1)}}{2}\right) -$$

$$\left. - \mu^2\frac{B^{(1)}}{2}a_1 - \frac{B^{(2)}}{4}\bar{l}_{FH}a_1 - \frac{B^{(3)}}{6}\bar{\Omega}_x - \frac{\lambda_{b1}^{(2)}}{4}\right\},$$

$$C_{my} := \frac{M_{HUBy}}{\frac{1}{2}\rho\pi R^3(\omega R)^2} = \bar{l}_{FH}\sigma\bar{C}_L^{\alpha}\left\{\frac{JR}{2\gamma I_{zz}^{(r)}}\left[a_1 - \left(1+\frac{m_B}{J}l_{FH}\right)\left(2\bar{\Omega}_x + \bar{\varepsilon}_{\Omega y}\right)\right] + \right.$$

$$+ \left(\frac{B^{(3)}}{6} + \mu^2\frac{B^{(1)}}{8}\right)(\theta_1 - b_1 - ka_1) +$$

$$\left. + \mu\frac{B^{(2)}}{4}a_0 + \frac{B^{(2)}}{4}\bar{l}_{FH}b_1 - \frac{B^{(3)}}{6}\bar{\Omega}_y - \frac{\lambda_{a1}^{(2)}}{4}\right\}.$$

$$(8.7)$$

There mathematical expressions represent calculation of the lateral and longitudinal hub moments at specified rotor conditions in terms of the characteristic measures of non-uniform parameters.

8.2.6 The dominant component of the lateral and longitudinal hub moments is the moment caused by accelerated flapping motion of rotor blades at blade cone tilt, as was described in 8.2.3. The moment acts toward the blade cone tilt, that can be interpreted as a moment caused by the cone tilt, which is a result of the swashplate tilt. Only these components determine the lateral and longitudinal hub moments at hovering or vertical motion ($\mu = 0$) without rotation of the rotor system ($\Omega_x = 0$, $\Omega_y = 0$) and neglecting the second power of the normalized flapping hinge offset ($\bar{l}_{FH}^2 \approx 0$) with taking into account that $b_1 + ka_1 - \theta_1 \approx 0$ and $a_1 - kb_1 + \theta_2 \approx 0$ (7.8) at such conditions:

$$M_{HUBx} = \bar{l}_{FH}\frac{nJ\omega^2}{2}b_1,$$

$$M_{HUBy} = \bar{l}_{FH}\frac{nJ\omega^2}{2}a_1.$$

This gives possibility to perform control over the rotorcraft via the swashplate mechanism together with the longitudinal and side rotor forces. These dominant components of the longitudinal and lateral hub moments correspond to the conventional blade element rotor theory.

The superposition of the lateral and longitudinal hub moments responds to the rotor system rotation (Ω_x, Ω_y) by a damp response and by a transverse response.

The damp response is caused by change of attack angles of blade elements due to circumferential element velocities, which are caused by the rotor system rotation. This causes azimuthally inhomogeneous blade lift forces, which is transferred on flapping hinges as azimuthally inhomogeneous vertical forces on the hinges, which cause a response moment as superposition the correspondent lateral and longitudinal moments. This response moment act in opposite direction to the rotor system rotation, is proportional to the angular velocity of this rotation, and is determined by aerodynamic properties of the rotor blades. This response is represented by the term with Ω_x in the expression of M_{HUBx} and by the term with Ω_y in the expression of M_{HUBy} (8.6).

The transverse response is a superposition of the lateral and longitudinal moments caused by Coriolis forces due to the rotor system rotation accompanied by the rotation around the rotor shaft, as was described in 8.2.3 at discussion about the moment on the hub caused by inertial forces. This moment acts in azimuthal direction which is $\pi/2$ ahead from direction of this rotor system rotation ($M_{HUBy} \propto -\Omega_x$, $M_{HUBx} \propto \Omega_y$), i.e transversely to this rotation.

The azimuthally inhomogeneous blade lift forces appear at forward motion, which affect the lateral and longitudinal moments. Different air velocity of advancing and retreating blades causes lateral moment toward $3\pi/2$ azimuthal position ($M_{HUBx} \propto -\mu(\theta_0 + \varphi^{(2)} - ka_0)$); different blowing of front and back blades causes longitudinal moment proportionally a_0 toward zero azimuthal position ($M_{HUBy} \propto \mu a_0$). The azimuthal inhomogeneity of the induced airflow, which appears at forward motion ($\lambda_{a1}^{(2)} \neq 0$, $\lambda_{b1}^{(2)} \neq 0$), contributes to these moments as well.

8.3 BLADE RESISTANCE MOMENT ON THE ROTOR HUB

8.3.1 The component M_{HUBz} of the rotor hub moment (8.2c) represents the rotor resistance moment, which is created by rotor blades interacting with their flapping hinges and acts on the rotor hub around the z_r-axis of the rotor frame, which coincides with the rotor axis. Generally, the moment acts in opposite direction to the rotation around the rotor shaft. The moment damps the angular speed of this rotation due to friction of the blades with air, as well as due to induced drag, which is the result of generation of blade lift forces. Some torque is necessary to be applied to the rotor shaft to compensate for the blade resistance moment in order to keep the shaft rotation; such torque is usually produced by an engine of the rotorcraft.

This resistance moment directly affects the rotor hub and the rotor shaft; however, this moment is not explicitly transferred on the rotorcraft. The rotor shaft does not interact with the rotorcraft construction exclusively only at the rotation around the rotor axis except interaction with an engine solidly fixed to the rotorcraft. As it is described in section 8.4, the resistance moment on the hub implicitly affects the rotorcraft via the interaction of the rotor shaft with the engine.

8.3.2 The hub moment M_{HUBz} may act in same direction as the shaft rotation at certain rotor conditions when the air blows from below of the rotor disc through the disc upward of the rotor system. Such conditions appear at fast descent of the rotorcraft. This causes slight tilt of element lift forces in direction of circumferential motion around the shaft; the projections of these lift forces on the tangential axis are directed toward this motion and promote the rotation around the rotor shaft. The induced element drag is negative in such a case. These forces create a moment on the rotor hub around the rotor shaft which accelerates the shaft rotation at such conditions; this mode of the rotor system is called autorotation. This specific case is not considered here and remains out of the current discussion.

8.3.3   Before analysis of the resistance moment on the hub presented below, it is analyzed here the term $F_{FHy}^{(r)} \cos \psi + F_{FHx}^{(r)} \sin \psi$, which is a part of the integrand in the M_{HUBz} expression (8.2c). This term represents a tangential component F_{FHx} (along the tangential x-axis) of a blade force on a flapping hinge in the blade-hinge frame, which was defined in 6.2.1:

$$F_{FHx} = F_{FHy}^{(r)} \cos \psi + F_{FHx}^{(r)} \sin \psi.$$

Positive direction of this force coincides with direction of the circumferential motion of the flapping hinge around the rotor shaft; the force component lays in the rotor plane and is applied to the blade flapping hinge. This force creates a moment on the rotor hub around the rotor axis with an arm of the flapping hinge offset l_{FH} (fig. 8.1).

As was shown in 6.2.1, this tangential component is opposite to the reaction force of the slip stopper of the flapping hinge R_{slip}, which prevents slipping of the blade lugs along the hinge axle: $F_{FHx} = -R_{slip}$. The slip stopper reaction force R_{slip} was found as a solution of the blade motion equation system (6.2), according to which:

$$F_{FHx} = -R_{slip} = \int\limits_{0}^{L_B} dX - \frac{d\omega}{dt}(J\cos\beta + m_B l_{FH}) - J\sin\beta(\varepsilon_{\Omega x}\cos\psi - \varepsilon_{\Omega y}\sin\psi) +$$

$$+ 2J\frac{d\beta}{dt}\omega\sin\beta - 2J\frac{d\beta}{dt}\cos\beta(\Omega_x\cos\psi - \Omega_y\sin\psi) +$$

$$+ (J\cos\beta + m_B l_{FH})(\Omega_x\cos\psi - \Omega_y\sin\psi)(\Omega_x\sin\psi + \Omega_y\cos\psi)$$

The first summand represents the blade aerodynamic force acting on the hinge in the tangential direction and consists of profile drag and induced drag of the blade. All other summands represent inertial forces of the whole blade. The second summand represents inertial forces due to accelerated rotation of the blade around the rotor shaft. The third summand represents a projection on the x-tangential axis at angle $\pi/2 - \beta$ of inertial forces due to accelerated rotation of the whole rotor system. The fourth summand represents a projection on the x-axis at angle $\pi/2 - \beta$ of Coriolis forces due to rotation around the rotor shaft accompanied by the flapping motion. The fifth summand represents Coriolis forces due to rotation of the rotor system accompanied by the flapping motion. The sixth summand is centrifugal forces due to rotor system rotation; this term can be neglected due to small magnitudes of angular velocities of the rotor system rotation ($|\Omega_x| \ll |\omega|$, $|\Omega_x| \ll |\omega|$). The tangential component of the blade force on the hinge can be expressed in a simplified form without the neglected terms and with the assumption of small flapping angle that

$\sin\beta \approx \beta$ and $\cos\beta \approx 1$:

$$F_{FHx} = \int_0^{L_B} dX - \frac{d\omega}{dt}(J + m_B l_{FH}) - J\omega^2\beta(\bar{\varepsilon}_{\Omega x}\cos\psi - \bar{\varepsilon}_{\Omega y}\sin\psi) + \tag{8.8}$$

$$+ 2J\omega\frac{d\beta}{dt}(\beta - \bar{\Omega}_x\cos\psi + \bar{\Omega}_y\sin\psi).$$

8.3.4   The resistance moment on the hub M_{HUBz} (8.2c) is analyzed on the base of the found tangential component of the blade force on the flapping hinge (8.8) and the blade moment on the hinge M_{onFH} according to (5.29):

$$M_{HUBz} = -l_{FH}F_{FHz} + M_{onFH} = -\frac{n}{2\pi}\int_0^{2\pi}\int_0^{L_B}(y_b + l_{FH})dXdy_b d\psi +$$

$$+ \frac{n}{2\pi}\int_0^{2\pi}\left[\left(I_{zz}^{(r)} + l_{FH}(J + m_B l_{FH})\right)\frac{d\omega}{dt} + \right. \tag{8.9}$$

$$+ \omega^2\beta(I_{zz}^{(b)} + l_{FH}J)(\bar{\varepsilon}_{\Omega x}\cos\psi - \bar{\varepsilon}_{\Omega y}\sin\psi) -$$

$$\left. - 2\omega\frac{d\beta}{dt}(I_{Cyy}^{(b)} + l_{FH}J)(\beta - \bar{\Omega}_x\cos\psi + \bar{\Omega}_y\sin\psi)\right]d\psi.$$

This expression determines the resistance moment on the hub in terms of blade aerodynamic drag and the blade inertial forces.

The second integral of (8.9) with the inertial forces is found with the assumption of flapping motion limited by first harmonic (5.12):

$$M_{HUBz} = -\frac{n}{2\pi}\int_0^{2\pi}\int_0^{L_B}(y_b + l_{FH})dXdy_b d\psi +$$

$$+ n\left(I_{zz}^{(r)} + l_{FH}(J + m_B l_{FH})\right)\frac{d\omega}{dt} + \frac{n\omega^2}{2}(I_{zz}^{(b)} + Jl_{FH})(\bar{\varepsilon}_{\Omega y}b_1 - \bar{\varepsilon}_{\Omega x}a_1) -$$

$$- n\omega^2(I_{Cyy}^{(b)} + l_{FH}J)(b_1\bar{\Omega}_x + a_1\bar{\Omega}_y).$$

The solved term with the inertial forces (the second, third, and fourth summands) does not fully represent impact of the blade inertial forces; as it will be shown below, there is implicit contribution of the inertial force arising from an interpretation of the aerodynamic part of the M_{HUBz}.

8.3.5   The dominant part of the resistance moment on the hub is the moment created by the aerodynamic tangential forces of all blades (dX). These forces create the following moment components: the blade moment on the blade flapping hinge (M_{onFH}), caused by separate location of blade lugs on the flapping hinge axle; and the moment created by tangential blade force (F_{FHx}) on the hinge with the arm of flapping hinge offset (l_{FH}). The first integral of the M_{HUBz} expression (8.9) represents this part due to these aerodynamic forces. This integral can be expressed with the extended tangential component of an element aerodynamic force (dX) according to (3.17):

$$M_{HUBz} = -\frac{\rho \omega R^2}{2} \frac{n}{2\pi} \int_0^{2\pi} \int_0^{L_B} c(y_b) C_L^\alpha(y_b) \omega(y_b + l_{FH}) \bar{V}_{ez} (\varphi_e \bar{V}_{ex} + \bar{V}_{ez}) dy_b d\psi +$$

$$+ \frac{\rho(\omega R)^2}{2} \frac{n}{2\pi} \int_0^{2\pi} \int_0^{L_B} c(y_b) C_D(y_b) (y_b + l_{FH}) \bar{V}_{ex}^2 dy_b d\psi +$$

$$+ n \left(I_{zz}^{(r)} + l_{FH}(J + m_B l_{FH}) \right) \frac{d\omega}{dt} + \frac{n\omega^2}{2} (I_{zz}^{(b)} + Jl_{FH})(\bar{\varepsilon}_{\Omega y} b_1 - \bar{\varepsilon}_{\Omega x} a_1) -$$

$$- n\omega^2 (I_{Cyy}^{(b)} + l_{FH}J)(b_1 \bar{\Omega}_x + a_1 \bar{\Omega}_y),$$

where the first summand represents a resistance moment caused by induced drag of all blade elements of all blades, and the second summand represents a resistance moment due to profile drag of all blade elements of all blades.

Some manipulations are performed in order to explain the structure of the resistance moment. The multiplier $\omega(y_b + l_{FH})$ in the integrand of the induced drag integral (the first summand) is considered here as the circumferential velocity of blade element rotation around the rotor shaft and is part of the tangential coordinate of element air velocity $\bar{V}_{ex}$ (3.6):

$$\omega(y_b + l_{FH}) = (\omega R)\bar{V}_{ex} - (\omega R)\mu \sin \psi.$$

The expression for the M_{HUBz} can be rewritten with such interpretation, with the extension only the normal coordinate of element air velocity $\bar{V}_{ez}$ in the first integral according to (3.6), with the assumption that flapping motion is limited by first harmonics (5.12), and with following sorting:

$$M_{HUBz} = -R \frac{\rho \omega R^2}{2} \frac{n}{2\pi} \int_0^{2\pi} \int_0^{L_B} \lambda(y, \psi) c(y_b) C_L^\alpha(y_b) \bar{V}_{ex}(\varphi_e \bar{V}_{ex} + \bar{V}_{ez}) dy_b d\psi +$$

$$+ R\mu \frac{\rho \omega R^2}{2} \frac{n}{2\pi} \int_0^{2\pi} \int_0^{L_B} c(y_b) C_L^\alpha(y_b)(\bar{V}_{ex}\beta \cos \psi + \bar{V}_{ez} \sin \psi)(\varphi_e \bar{V}_{ex} + \bar{V}_{ez}) dy_b d\psi +$$

$$+ (a_1 - \bar{\Omega}_x) \frac{\rho \omega R^2}{2} \frac{n}{2\pi} \int_0^{2\pi} \int_0^{L_B} c(y_b) C_L^\alpha(y_b) \bar{V}_{ex}(\varphi_e \bar{V}_{ex} + \bar{V}_{ez}) y_b \sin \psi dy_b d\psi -$$

$$- (b_1 + \bar{\Omega}_y) \frac{\rho \omega R^2}{2} \frac{n}{2\pi} \int_0^{2\pi} \int_0^{L_B} c(y_b) C_L^\alpha(y_b) \bar{V}_{ex}(\varphi_e \bar{V}_{ex} + \bar{V}_{ez}) y_b \cos \psi dy_b d\psi -$$

$$- l_{FH}\bar{\Omega}_x \frac{\rho \omega R^2}{2} \frac{n}{2\pi} \int_0^{2\pi} \int_0^{L_B} c(y_b) C_L^\alpha(y_b) \bar{V}_{ex}(\varphi_e \bar{V}_{ex} + \bar{V}_{ez}) \sin \psi dy_b d\psi -$$

$$- l_{FH}\bar{\Omega}_y \frac{\rho \omega R^2}{2} \frac{n}{2\pi} \int_0^{2\pi} \int_0^{L_B} c(y_b) C_L^\alpha(y_b) \bar{V}_{ex}(\varphi_e \bar{V}_{ex} + \bar{V}_{ez}) \cos \psi dy_b d\psi +$$

$$+ \frac{\rho(\omega R)^2}{2} \frac{n}{2\pi} \int_0^{2\pi} \int_0^{L_B} c(y_b) C_D(y_b)(y_b + l_{FH}) \bar{V}_{ex}^2 dy_b d\psi +$$

$$+ n(I_{zz}^{(r)} + l_{FH}(J + m_B l_{FH})) \frac{d\omega}{dt} + \frac{n\omega^2}{2}(I_{zz}^{(b)} + Jl_{FH})(\bar{\varepsilon}_{\Omega y} b_1 - \bar{\varepsilon}_{\Omega x} a_1) -$$

$$- n\omega^2 (I_{Cyy}^{(b)} + l_{FH}J)(b_1 \bar{\Omega}_x + a_1 \bar{\Omega}_y).$$

This huge expression can be simplified with the determination of the blade normal component dZ according to (3.17), which represents blade element lift force as:

$$dZ = -\frac{1}{2}\rho c(y_b)(\omega R)^2 C_L^\alpha(y_b)\left(\varphi_e \bar{V}_{ex}^2 + \bar{V}_{ex}\bar{V}_{ez}\right)dy_b.$$

Besides this, the second summand represents a part of the rotor longitudinal force H, which is caused by blade lift forces, as was discussed in section 7.3. This summand can be represented in terms of the rotor longitudinal force according to (7.4) with $\sin\beta \approx \beta$ and $\cos\beta \approx 1$:

$$\frac{\rho(\omega R)^2}{2}\frac{n}{2\pi}\int_0^{2\pi}\int_0^{L_B}c(y_b)C_L^\alpha(y_b)\left(\varphi_e\bar{V}_{ex}+\bar{V}_{ez}\right)\left(\bar{V}_{ex}\beta\cos\psi+\bar{V}_{ez}\sin\psi\right)dy_bd\psi =$$

$$= -H + \frac{\rho(\omega R)^2}{2}\frac{n}{2\pi}\int_0^{2\pi}\int_0^{L_B}c(y_b)C_D(y_b)\bar{V}_{ex}\left(\bar{V}_{ex}\sin\psi - \bar{V}_{ey}\cos\psi\right)dy_bd\psi.$$

Implementing these, the simplified expression has the view:

$$M_{HUBz} = R\frac{n}{2\pi}\int_0^{2\pi}\int_0^{L_B}\lambda(y,\psi)dZd\psi - R\mu H -$$

$$-(a_1 - \bar{\Omega}_x)\frac{n}{2\pi}\int_0^{2\pi}\left[\int_0^{L_B}y_bdZ\right]\sin\psi d\psi +$$

$$+(b_1 + \bar{\Omega}_y)\frac{n}{2\pi}\int_0^{2\pi}\left[\int_0^{L_B}y_bdZ\right]\cos\psi d\psi +$$

$$+\bar{\Omega}_x l_{FH}\frac{n}{2\pi}\int_0^{2\pi}\int_0^{L_B}dZ\sin\psi d\psi +$$

$$+\bar{\Omega}_y l_{FH}\frac{n}{2\pi}\int_0^{2\pi}\int_0^{L_B}dZ\cos\psi d\psi + \qquad (8.10)$$

$$+\frac{\rho(\omega R)^2}{2}\frac{n}{2\pi}\int_0^{2\pi}\int_0^{L_B}c(y_b)C_D(y_b)\left[(y_b+l_{FH})\bar{V}_{ex}^2 + \right.$$

$$\left. + R\mu\bar{V}_{ex}\left(\bar{V}_{ex}\sin\psi - \bar{V}_{ey}\cos\psi\right)\right]dy_bd\psi +$$

$$+n\left(I_{zz}^{(r)}+l_{FH}(J+m_Bl_{FH})\right)\frac{d\omega}{dt} + \frac{n\omega^2}{2}(I_{zz}^{(b)}+Jl_{FH})(\bar{\varepsilon}_{\Omega y}b_1 - \bar{\varepsilon}_{\Omega x}a_1) -$$

$$-n\omega^2(I_{Cyy}^{(b)}+l_{FH}J)(b_1\bar{\Omega}_x+a_1\bar{\Omega}_y).$$

The integral $\int_0^{L_B}y_bdZ$ corresponds to the moment of blade lift forces around the blade flapping hinge. This moment participates in the flapping motion and can be found from the blade flapping equation (5.9):

$$\int_0^{L_B}y_bdZ = \omega^2 I_{zz}^{(r)}\left(-\epsilon\frac{1}{\omega^2}\frac{d^2\beta}{dt^2} - \beta + \left(2\bar{\Omega}_x+\bar{\varepsilon}_{\Omega y}\right)\cos\psi - \left(2\bar{\Omega}_y - \bar{\varepsilon}_{\Omega x}\right)\sin\psi\right).$$

Assuming that flapping motion is limited by first harmonics (5.12):

$$\int_0^{L_B} y_b dZ = \omega^2 I_{zz}^{(r)} \Big(-a_0 + \big[(1-\epsilon)a_1 + 2\bar{\Omega}_x + \bar{\epsilon}_{\Omega y}\big]\cos\psi +$$

$$+ \big[(1-\epsilon)b_1 - 2\bar{\Omega}_y + \bar{\epsilon}_{\Omega x}\big]\sin\psi\Big).$$

The summands of (8.10), which contain this integral, represent the impact of inertial forces, which appear at flapping motion. These summands are expressed with this integral:

$$-(a_1 - \bar{\Omega}_x)\frac{n}{2\pi}\int_0^{2\pi}\left[\int_0^{L_B} y_b dZ\right]\sin\psi\, d\psi + (b_1 + \bar{\Omega}_y)\frac{n}{2\pi}\int_0^{2\pi}\left[\int_0^{L_B} y_b dZ\right]\cos\psi\, d\psi =$$

$$= \frac{1}{2}n\omega^2 I_{zz}^{(r)}\Big(-(a_1 - \Omega_x)\bar{\epsilon}_{\Omega x} + (b_1 + \Omega_y)\bar{\epsilon}_{\Omega y} + (3-\epsilon)(\Omega_x b_1 + \Omega_y a_1)\Big),$$

where $I_{zz}^{(r)} = I_{zz}^{(b)} + Jl_{FH}$ and $\epsilon = I_{xx}^{(b)}/I_{zz}^{(r)}$ are introduced in 5.2.1 and 5.2.2.

The integrals $(l_{FH}\frac{n}{2\pi}\int_0^{2\pi}\int_0^{L_B} dZ\sin\psi\, d\psi)$ and $(l_{FH}\frac{n}{2\pi}\int_0^{2\pi}\int_0^{L_B} dZ\cos\psi\, d\psi)$ represent the aerodynamic parts of the correspondent longitudinal M_{HUBx} and lateral M_{HUBy} hub moments according to (8.5) and can be expressed in terms of these moments excluding the parts caused by inertial forces:

$$l_{FH}\frac{n}{2\pi}\int_0^{2\pi}\int_0^{L_B} dZ\sin\psi\, dy_b d\psi =$$

$$= M_{HUBx} - l_{FH}\frac{nJ\omega^2}{2}\left(b_1 + \left[1 + \frac{m_B}{J}l_{FH}\right](2\bar{\Omega}_y - \bar{\epsilon}_{\Omega x})\right),$$

$$l_{FH}\frac{n}{2\pi}\int_0^{2\pi}\int_0^{L_B} dZ\cos\psi\, dy_b d\psi =$$

$$= M_{HUBy} - l_{FH}\frac{nJ\omega^2}{2}\left[a_1 - \left(1 + \frac{m_B}{J}l_{FH}\right)(2\bar{\Omega}_x + \bar{\epsilon}_{\Omega y})\right].$$

Applying all of the specified into the expression (8.10), the resistance moment can be expressed in the following way:

$$M_{HUBz} = R\frac{n}{2\pi} \int_0^{2\pi} \int_0^{L_B} \lambda(y,\psi)dZd\psi - R\mu H + \bar{\Omega}_x M_{HUBx} + \bar{\Omega}_y M_{HUBy} +$$

$$+ \frac{\rho(\omega R)^2 RnS_B}{2} \left(\frac{\bar{C}_D^{(4)}}{4} + \mu^2 \bar{C}_D^{(2)} \right) +$$

$$+ n\left(I_{zz}^{(r)} + l_{FH}(J + m_B l_{FH})\right) \left\{ \frac{d\omega}{dt} + \omega^2 \left[\left(\frac{\bar{\Omega}_x}{2} - a_1 \right) \bar{\varepsilon}_{\Omega x} + \left(\frac{\bar{\Omega}_y}{2} + b_1 \right) \bar{\varepsilon}_{\Omega y} \right] \right\} -$$

$$- n\omega^2 l_{FH}(J + m_B l_{FH})(b_1 \bar{\varepsilon}_{\Omega y} - a_1 \bar{\varepsilon}_{\Omega x}) +$$

$$+ n\frac{\omega^2}{2} \left(I_{zz}^{(b)}(1-\epsilon) - J l_{FH}\epsilon - 2(I_{Cyy}^{(b)} - I_{zz}^{(b)}) \right) (\bar{\Omega}_x b_1 + \bar{\Omega}_y a_1).$$

$$(8.11)$$

The term $I_{Cyy}^{(b)} - I_{zz}^{(b)}$ is negligibly small due to the high wing aspect ratio, for which $I_{Cyy}^{(b)} \approx I_{zz}^{(b)}$; and this term can be neglected. The ratio ϵ equals to $I_{zz}^{(b)}/(I_{zz}^{(b)} + J l_{FH})$ assuming $I_{zz}^{(b)} \approx I_{xx}^{(b)}$; therefore, the last summand with these terms is reduced to zero:

$$n\frac{\omega^2}{2} \left(I_{zz}^{(b)}(1-\epsilon) - J l_{FH}\epsilon - 2(I_{Cyy}^{(b)} - I_{zz}^{(b)}) \right) (\bar{\Omega}_x b_1 + \bar{\Omega}_y a_1) \approx 0.$$

8.3.6   According to the found expression (8.11), the resistance moment on the rotor hub is caused by the following components.

- The summand $R\frac{n}{2\pi} \int_0^{2\pi} \lambda(y,\psi)dZd\psi$ represents a component, which is created by element induced drag caused by passing the airflow through the rotor. This resistance moment component is interpreted as a result of the rotor thrust generation.
- The summand $-R\mu H$ represents a resistance moment component created by element induced drag caused by forward motion of the rotor system at acting of longitudinal rotor force towards this motion $(-H)$. This component is interpreted as a result of generation of the longitudinal rotor force causing forward motion.
- The summand $\bar{\Omega}_x M_{HUBx}$ represents a resistance moment component created by element induced drag caused by lateral rotation of the rotor system under the lateral hub moment. This component is a result of the lateral rotation of the rotor system under the lateral hub moment.
- The summand $\bar{\Omega}_x M_{HUBy}$ represents a resistance moment component created by element induced drag caused by longitudinal rotation of the rotor system under the longitudinal hub moment. This component is a result of the longitudinal rotation under the longitudinal hub moment.
- The summand $\frac{\rho(\omega R)^2 RnS_B}{8} \left(\bar{C}_D^{(4)} + 4\mu^2 \bar{C}_D^{(2)} \right)$ represents a component of the resistance moment created by element profile drag caused by friction of the blades with air.

− The summands with $d\omega/dt$, $\bar{\varepsilon}_{\Omega x}$, and $\bar{\varepsilon}_{\Omega y}$ represent components of the resistance moment, which are caused by inertial resistance to change of the angular velocities of the rotor blades rotation around the rotor shaft and the rotor system rotation.

8.3.7 The rotor system consumes energy during the rotation around the rotor shaft (at no autorotation). The consumption energy of the rotor system is defined here as the energy, which is spent to overcome the profile drag of the blades caused by friction of the blades with air and the induced drag caused by the generation of lift forces. The consumption energy is represented by the resistance moment on the hub, which is transferred on the rotor shaft at certain turn of the rotor blades around the rotor shaft. However, it is useful to operate with a consumption power of the rotor system, which represents the consumption energy during a certain period unit of time. This rotor consumption power $P_{consume}$ is determined by a product of the resistance moment on the hub M_{HUBz} and the angular speed of the shaft rotation together with the hub ω:

$$P_{consume} = \omega M_{HUBz}.$$

The consumption power is considered to be positive if the rotor system requires external torque directed toward the rotor shaft rotation in order to compensate for the resistance moment and to keep the rotation speed constant ($M_{HUBz} > 0$); therefore, there is a minus sign. The consumption power can be decomposed into the following components according to the resistance moment components (8.11):

$$P_{consume} = \frac{n}{2\pi} \int_0^{2\pi} \int_0^{L_B} (V_y + V_i(y, \psi))\, dZ d\psi + V_x(-H) +$$

$$+ \Omega_x M_{HUBx} + \Omega_y M_{HUBy} + \frac{\rho(\omega R)^3 (\pi R^2)\sigma}{2} \left(\frac{\bar{C}_D^{(4)}}{4} + \mu^2 \bar{C}_D^{(2)} \right) +$$

$$+ n\omega \left(I_{zz}^{(r)} + l_{FH}(J + m_B l_{FH}) \right) \left\{ \frac{d\omega}{dt} + \omega^2 \left[\left(\frac{\bar{\Omega}_x}{2} - a_1 \right) \bar{\varepsilon}_{\Omega x} + \left(\frac{\bar{\Omega}_y}{2} + b_1 \right) \bar{\varepsilon}_{\Omega y} \right] \right\} -$$

$$- n\omega^3 l_{FH}(J + m_B l_{FH})(b_1 \bar{\varepsilon}_{\Omega y} - a_1 \bar{\varepsilon}_{\Omega x}).$$

$$(8.12)$$

The consumption power consists of the following components.

The first summand (8.12) represents power, which is consumed by the rotor system in order to generate blade lift forces by passing airflow through the rotor disc. The passing airflow consists of the induced airflow $V_i(y, \psi)$ and airflow caused by vertical motion of the rotor system V_y. This summand is interpreted as a power that is required to generate the thrust, which represents the total lift force of all blades. This power component is called the induced power of the rotor system. If the rotor system moves upward ($V_y < 0$) and elements of the rotor blades create lift forces directed upward with respect to the rotor system ($dZ < 0$), then the rotor system creates induced airflow, which passes the rotor disc from above of the rotor disc downward, so the rotor system moves upward relative to this induced airflow ($V_i(y, \psi) < 0$); therefore, the induced power is positive.

The second summand represents power, which is consumed by the rotor system in order to generate rotor force $(-H)$ toward forward motion (V_x) of the rotor system. This component represents the power, which is consumed to move the rotor system in the forward direction (along V_x). If the longitudinal rotor force H is directed toward the forward motion, then this force is negative; therefore, this power component is positive.

The third and fourth summands represent power, which the rotor system consumes to provide the rotor system rotation. These components together equal the hub moment multiplied on the angular velocity of the rotor system rotation, that is equivalent to the sum of multiplications of the lateral hub moment on lateral angular velocity and the longitudinal hub moment on longitudinal angular velocity of the rotor system rotation.

The fifth summand represents power, which is consumed by friction of blade elements with air (blade element profile drag) during rotation around the rotor shaft and is called the profile power. This power component increases at forward motion because of increase of air velocities of the blade elements. The profile power is always positive.

The last two summands in (8.12) represent power, which is consumed by the rotor system to accelerate the rotor shaft rotation and the rotor system rotation.

8.3.8   A coefficient of the resistance moment on the hub is introduced in order to generalize the resistance moment over rotor system size, rotor shaft angular speed, and air density. The resistance moment coefficient C_{mz} is defined as the resistance moment on the hub normalized to the rotor disc area (πR^2), to dynamic pressure of airflow on a blade tip at circumferential motion $(\rho(\omega R)^2/2)$, and to the rotor disc radius (R) as the rotor system reference length:

$$C_{mz} := \frac{M_{HUBz}}{\frac{1}{2}\rho\pi R^3(\omega R)^2} = -\frac{1}{2\pi}\int_0^{2\pi}\int_0^{L_B}\lambda(y,\psi)dC_T d\psi - \mu C_H +$$

$$+ \sigma\left(\frac{\bar{C}_D^{(4)}}{4} + \mu^2\bar{C}_D^{(2)}\right) + \bar{\Omega}_x C_{mx} + \bar{\Omega}_y C_{my} +$$

$$+ n\frac{I_{zz}^{(r)} + l_{FH}(J + m_B l_{FH})}{\frac{1}{2}\rho\pi R^3(\omega R)^2}\left\{\frac{d\omega}{dt} + \omega^2\left[\left(\frac{\bar{\Omega}_x}{2} - a_1\right)\bar{\varepsilon}_{\Omega x} + \left(\frac{\bar{\Omega}_y}{2} + b_1\right)\bar{\varepsilon}_{\Omega y}\right]\right\} -$$

$$- l_{FH}n\frac{J + m_B l_{FH}}{\frac{1}{2}\rho\pi R^5}(b_1\bar{\varepsilon}_{\Omega y} - a_1\bar{\varepsilon}_{\Omega x}),$$

$$(8.13)$$

where: $dC_T = -\dfrac{ndZ}{\frac{1}{2}\rho\pi R^2(\omega R)^2}$.

The term dC_T represents a part of generalized lift forces distributed around the rotor disc, which acts per segment of the disc circle $d\psi$ with width dy_b.

In correspondence to the resistance moment coefficient, a coefficient of the consumption power is introduced. The consumption power coefficient equals to the consumption power normalized to the rotor disc area (πR^2), to airflow dynamic pressure of airflow on a blade tip at circumferential motion ($\rho(\omega R)^2/2$), to the rotor disc radius (R), and to the angular speed of the rotor shaft rotation (ω). Essentially, such consumption power coefficient equals the resistance moment coefficient:

$$C_{Pconsume} := \frac{P_{consume}}{\frac{1}{2}\rho\pi R^2(\omega R)^3} = C_{mz}. \tag{8.14}$$

The coefficients C_{mz} and $C_{Pconsume}$ are not used in the conventional blade element rotor theory: these are introduced here in order of consistent explanation.

8.3.9 An attention is paid here to the induced power of the rotor system, which is the dominant part of the rotor system consumption power. The induced power can be represented in terms of characteristic measures introduced in section 4:

$$-\frac{1}{2\pi}\int_0^{2\pi}\int_0^{L_B}\lambda(y,\psi)dC_Td\psi =$$

$$-(\theta_0-ka_0)\left(\frac{\lambda_0^{(3)}}{3}+\mu^2\frac{\lambda_0^{(1)}}{2}+\mu\frac{\lambda_{b1}^{(2)}}{2}\right)-\frac{\lambda_0^{tw(3)}}{3}-\mu^2\frac{\lambda_0^{tw(1)}}{2}-\mu\frac{\lambda_{b1}^{tw(2)}}{2}+$$

$$+\mu a_0\frac{\lambda_{a1}^{(2)}}{4}-\mu a_1\left(\frac{\lambda_0^{(2)}}{2}+\mu\frac{\lambda_{b1}^{(1)}}{2}\right)+$$

$$+(\theta_1-b_1-ka_1)\left(\frac{\lambda_{a1}^{(3)}}{6}+\mu^2\frac{\lambda_{a1}^{(1)}}{8}\right)+$$

$$+(\theta_2+a_1-kb_1)\left(\mu\frac{\lambda_0^{(2)}}{2}+\frac{\lambda_{b1}^{(3)}}{6}+3\mu^2\frac{\lambda_{b1}^{(1)}}{8}\right)-$$

$$-\Omega_x\left(\mu\frac{\lambda_0^{(2)}}{4}+\frac{\lambda_{b1}^{(3)}}{6}\right)-\Omega_y\frac{\lambda_{a1}^{(3)}}{6}+\bar{l}_{FH}\left(-a_1\left(\mu\frac{\lambda_0^{(1)}}{2}+\frac{\lambda_{b1}^{(2)}}{4}\right)+b_1\mu\frac{\lambda_{a1}^{(2)}}{4}\right)-$$

$$-\frac{1}{2}\Lambda_0^2-\frac{\mu}{2}\Lambda_b^2. \tag{8.15}$$

The induced power is positive at upward motion of the rotor system relative to oncoming airflow: the minus sign is compensated by the negative value of the inflow ratio $\lambda(y,\psi)$ at such upward motion. This induced power is described by the wide range of the inflow ratio characteristic parameters λs. The induced power depends on the induced airflow inhomogeneity around the rotor disc, as well as the blade twisting and the blade non-uniform parameters ($c(y_b)$, $C_L^{\alpha}(y_b)$).

Considering the case of homogeneous inflow around the rotor disc with the averaged inflow ratio λ_0 ($\lambda(y,\psi)=\lambda_0=const$), which is used in the conventional blade element rotor theory. This case specifies that:

$\lambda_0^{(n)} = B^{(n)} \lambda_0, \quad \lambda_0^{tw(n)} = B^{(n)} \varphi^{(n)} \lambda_0, \quad \Lambda_0^2 = B^{(2)} \lambda_0^2, \quad$ and

$\lambda_{a1}^{(n)} = 0, \quad \lambda_{b1}^{(n)} = 0, \quad \lambda_{a1}^{tw(n)} = 0, \quad \lambda_{b1}^{tw(n)} = 0, \quad \Lambda_b^2 = 0.$

The induced power at these conditions is reduced by an expression:

$$-\frac{1}{2\pi} \int_0^{2\pi} \int_0^{L_B} \lambda_0 dC_T d\psi = -\lambda_0 C_T|_{\lambda=\lambda_0},$$

where $C_T|_{\lambda=\lambda_0}$ is the rotor thrust coefficient at the homogeneous induced airflow:

$$C_T|_{\lambda=\lambda_0} = \sigma \bar{C}_L^\alpha \left[\frac{B^{(3)}}{3} \left(\theta_0 + \varphi^{(3)} - ka_0 \right) + \mu^2 \frac{B^{(1)}}{2} \left(\theta_0 + \varphi^{(1)} - ka_0 \right) + \frac{B^{(2)}}{2} \lambda_0^2 - \right.$$
$$\left. - \frac{B^{(2)}}{2} \mu \left(\theta_2 - kb_1 - a_1 \bar{l}_{FH} \frac{B^{(1)}}{B^{(2)}} - \frac{1}{2} \Omega_x \right) \right].$$

This corresponds to the conclusion of the conventional blade element rotor theory that proves the inheritance of the discussed theory to the conventional theory.

8.4 ROTOR TORQUE

8.4.1   An engine torque is usually applied to the rotor shaft in order to compensate for the resistance moment, which blades create on the rotor hub and the rotor shaft during their rotation. The engine torque is applied to the rotor shaft around the rotor axis (the z_r-axis of the rotor frame) in direction of the rotor shaft rotation and is denoted here by M_{engine}. The engine torque takes negative values since it acts toward the rotator shaft rotation, which was stated clockwise around the positive direction of the z_r-axis of the rotor frame. The engine torque provides constant rotation of the rotor shaft that ensures the permanent thrust generation for a continuous flight of the rotorcraft in the airspace. Engine torque applying can be interpreted in terms of energy conservation: the power, which is produced by the engine, is consumed by the rotor system.

Formally, engine interaction with the rotor shaft is beyond the discussed blade element rotor theory. The discussed model of the rotor system does not depend on an engine type and principles of engine design. However, the principles of interaction of the rotor system with the engine via the engine torque are important for helicopter aerodynamics.

8.4.2   The rotor shaft is rotated around the rotor axis under superposition of the resistance moment on the hub and an engine torque. The rotor shaft rotation around the rotor axis proceeds without any interaction with the rotorcraft construction except interaction with the engine in terms of the engine torque. If the engine torque compensates for the resistance moment on the hub ($M_{engine} = -M_{HUBz}$), then the rotation of the shaft is constant; if the engine torque dominates, then the shaft increases the angular speed of its rotation; if the resistance moment dominates, then the angular speed of the shaft rotation decreases.

The engine, which is solidly fixed to the rotorcraft, applies its torque to the rotor shaft. According to the principle of the Newton's third law, the rotor shaft applies force moment to the engine body, which equals by the magnitude to the engine torque but acts in opposite direction. This response of the rotor shaft on the engine body is called the reactive moment of the rotor system. Since the engine is solidly fixed to the rotorcraft, the reactive moment is transferred to the whole rotorcraft. So that, the rotorcraft is affected by the moment which is reversed to the engine torque applied to the rotor shaft. This reactive moment acting on the rotorcraft as a result of the engine torque application to the shaft is called the rotor torque and is denoted by Q. Formally, the rotor torque equals by magnitude and acts in opposite direction to the engine torque, and does not depend on the resistance moment on the hub.

8.4.3   In the case of the constant rotation of the rotor shaft ($d\omega/dt = 0$), the rotor torque must be the same by magnitude to the resistance moment on the hub but act in the opposite direction: $M_{engine} = -M_{HUBz}$. The rotor shaft affects the rotorcraft by the reactive torque (the rotor torque), which equals by the magnitude but acts in the opposite direction to the engine torque: $Q = -M_{engine}$. Therefore, the rotor torque has same magnitude and same direction (opposite to opposite) as the resistance moment:

$$Q = M_{HUBz}(\omega = const). \tag{8.16}$$

The condition of the constant rotor shaft rotation is typical and common for a normal rotorcraft (helicopter) flight when an engine operates properly. The rotor torque is commonly considered at this condition in the frame of the normal flight dynamics of a helicopter.

This case can be considered from the point of view of energy conservation. The engine releases a power P_{engine} during rotating the rotor shaft by applying the engine torque. This engine power is defined as a dot product of the engine torque aligned to the rotor axis $\vec{M}_{engine}(0, 0, M_{engine})$ and the rotor shaft angular velocity:

$$P_{engine} = \vec{\omega} \cdot \vec{M}_{engine}.$$

The engine power is considered to be positive when the engine torque is applied to the shaft in the direction of the shaft rotation. Since the rotor shaft rotates clockwise around the z_r-axis $\vec{\omega}(0, 0, -\omega)$, the engine torque must be negative ($M_{engine} < 0$); therefore, the engine power is positive. At the constant shaft rotation, the engine power is opposite to the resistance moment on the hub ($M_{engine} = -M_{HUBz}$):

$$P_{engine} = -\omega M_{engine} = \omega M_{HUBz}(\omega = const) = P_{consume}(\omega = const).$$

From this perspective, the generated engine power is spent on the consumption power needed to permanently rotate the blades around the rotor shaft.

8.4.4   A coefficient of the rotor torque C_Q is introduced here as the rotor torque normalized to the rotor disc area (πR^2), to dynamic pressure of airflow on a blade tip at circumferential motion ($\rho(\omega R)^2/2$), and to the rotor disc radius (R) as the rotor system reference length:

$$C_Q := \frac{Q}{\frac{1}{2}\rho \pi R^3 (\omega R)^2}. \tag{8.17}$$

This coefficient essentially equals to the resistance moment coefficient (8.13) at the constant shaft rotation:

$$C_Q = \frac{M_{HUBz}(\omega = const)}{\frac{1}{2}\rho\pi R^3 (\omega R)^2} = C_{mz}(\omega = const).$$

With this and with neglecting of rotor system accelerated rotations ($\bar{\varepsilon}_{\Omega x} \approx 0$ and $\bar{\varepsilon}_{\Omega y} \approx 0$), the rotor torque coefficient is expressed in a following way:

$$C_Q = -\frac{1}{2\pi} \int_0^{2\pi} \int_0^{L_B} \lambda(y,\psi) dC_T d\psi - \mu C_H +$$
$$+ \sigma \left(\frac{\bar{C}_D^{(4)}}{4} + \mu^2 \bar{C}_D^{(2)} \right) + \bar{\Omega}_x C_{mx} + \bar{\Omega}_y C_{my}. \qquad (8.18)$$

Correspondingly to the rotor torque coefficient, the blade element rotor theory operates with a rotor power coefficient C_P, which is defined as the engine power normalized to the rotor disc area (πR^2), to airflow dynamic pressure of airflow on a blade tip at circumferential motion ($\rho(\omega R)^2/2$), to the rotor disc radius (R), and to the angular speed of the rotor shaft (ω):

$$C_P := \frac{P_{engine}}{\frac{1}{2}\rho\pi R^2 (\omega R)^3} = C_Q. \qquad (8.19)$$

The rotor power coefficient generally equals the rotor torque coefficients and represents power, which is necessary to supply the rotor system in order to perform a rotorcraft flight. This is one of the important characteristics of helicopter flight dynamics.

9 Conclusions

In the retrospective of the blade element rotor theory presented above, the following conclusions are generalized about the dynamic properties of the rotor system with steady rotated blades.

9.1 DETERMINATION OF ROTOR SYSTEM PROPERTIES

9.1.1 The discussed blade element rotor theory describes the dynamic properties of the rotor system with the following parameters:

- Parameters of a blade cone, which the rotor blades circumscribe during the rotation around the rotor shaft accompanied by the blade flapping motion. These parameters consist of the cone angle a_0 and angles of the cone axis tilt a_1 and b_1 in the rotor longitudinal and lateral directions correspondingly.
- The total rotor force, which is applied to the hub center as the reference point. The force is decomposed into the rotor thrust T acting along the rotor axis upward, the longitudinal force H acting in opposite direction to the projection of the hub center velocity on the rotor plane, the side force S acting right-side with respect to the forward motion perpendicularly to T and H.
- The lateral M_{HUBx} and longitudinal M_{HUBy} moments on the hub, which is caused by the flapping hinge offset. The lateral moments M_{HUBx} acts around the rotor longitudinal x_r-axis; the longitudinal moment M_{HUBy} acts around the rotor y_r-lateral axis.
- The blade resistance moment M_{HUBz} on the hub, which is caused by the profile and induced drag of the rotor blades at no autorotation. This moment acts around the rotor axis.

The total rotor force, the lateral and longitudinal moments on the hub directly act on the rotorcraft, which the rotor system carries. The resistance moment explicitly does not act on the rotorcraft; however, the reactive torque acts on the rotorcraft, which is a reaction of the rotorcraft on an engine torque, that compensates for the resistance moment.

9.1.2 According to the discussed blade element rotor theory, these dynamic properties of the rotor system at steady blade rotation can be explicitly arithmetically calculated at known conditions of the rotor system, at known induced velocities in every part of the rotor disc, with a specified current state of the swashplate mechanism, with known parameters of the rotor system, and with a specified set of characteristic measures of blade non-uniform parameters.

More specifically, the following parameters need to be known for this calculation. The conditions of the rotor system mean the following set of parameters: the air density of the surrounded airspace ρ; the rotor shaft angular speed ω; velocity of the rotor system motion, which is described by velocity of the hub center specified in

DOI: 10.1201/9781003296232-9

the rotor frame by forward velocity V_x and vertical velocity V_z; the angular velocity of the rotor system, which is described by longitudinal angular velocity Ω_y around the y_r-axis of the rotor frame and lateral angular velocity Ω_x around the x_r-axis; the accelerated rotation of the rotor system is described by the derivatives of angular velocity components $\varepsilon_{\Omega x} = d\Omega_x/dt$ and $\varepsilon_{\Omega y} = d\Omega_y/dt$ with respect to time. The state of the swashplate mechanism is specified in terms of blade pitch (incidence) angle change by this mechanism and is described by the collective pitch θ_0, the lateral cyclic pitch angles θ_1 and the longitudinal cyclic pitch θ_2. The parameters of the rotor system remain constant and include the following parameters: the rotor radius R; the number of blades n; the flapping hinge offset l_{FH}; planform area of the rotor blade S_B; an average slope of the lift coefficient of the blade $\bar{C}_L^\alpha$; the blade mass m_B; the blade mass moment J; the blade moment of inertia around flapping hinge $I_{zz}^{(b)}$; the element of the Coriolis force tensor $I_{Cyy}^{(b)}$, which usually $I_{Cyy}^{(b)} \approx I_{zz}^{(b)}$; the flapping compensation coefficient k.

In order to calculate the parameters of the blade cone of the flapping motion limited by the first harmonic, it is enough to know the following sets of characteristic measures except those specified above parameters:

- the finite set of blade-hinge characteristic parameters $B_1^{(4)}, B_1^{(3)}, B_1^{(2)}, B_2^{(4)}$;
- the finite set of the twisting characteristic angles $\varphi_1^{(4)}, \varphi_1^{(3)}, \varphi_1^{(2)}$;
- the finite set of the inflow ratio characteristic parameters $\lambda_0^{(3,1)}, \lambda_{a1}^{(3,1)}, \lambda_{b1}^{(3,1)}, \lambda_0^{(2,1)}, \lambda_{a1}^{(2,1)}, \lambda_{b1}^{(2,1)}$.

The cone angle a_0 is calculated according to (5.25); the cone tilts a_1 and b_1 are calculated based on known a_0 according to (5.21).

The components of the total rotor force and of the rotor moment on the hub are calculated with the determined blade cone parameters and those parameters specified above. Besides this, it is enough to know the following sets of characteristic measures:

- the finite set of the blade characteristic parameters $B^{(1)}, B^{(2)}, B^{(3)}$;
- the finite set of the blade characteristic twisting angles $\varphi^{(1)}, \varphi^{(2)}, \varphi^{(3)}$;
- the finite set of the inflow ratio characteristic parameters $\lambda_0^{(3)}, \lambda_{a1}^{(3)}, \lambda_{b1}^{(3)}, \lambda_0^{(2)}, \lambda_{a1}^{(2)}, \lambda_{b1}^{(2)}, \lambda_0^{(1)}, \lambda_{a1}^{(1)}, \lambda_{b1}^{(1)}, \lambda_0^{tw(3)}, \lambda_0^{tw(1)}, \lambda_{a1}^{tw(2)}, \lambda_{b1}^{tw(2)}, \Lambda_0^2, \Lambda_a^2, \Lambda_b^2$;
- two characteristic drag coefficients $C_D^{(2)}, C_D^{(4)}$.

9.1.3 The components of the rotor force and rotor moment are calculated in terms of the correspondent coefficients: the thrust coefficient C_T according to (7.3); the longitudinal force coefficient C_H (7.7); the side force coefficient C_S (7.12); the lateral C_{mx} and longitudinal C_{my} hub moment coefficients (8.7); the resistance moment coefficient C_{mz} (8.11) with the induced power determined according to (8.15). The final components of the rotor force and rotor moment are calculated based on the found coefficients according to the definitions of these coefficients (7.2), (7.5), (7.10), and (8.7).

9.2 SUMMARIZING OF ROTOR SYSTEM PROPERTIES

9.2.1   The longitudinal axes of the blades circumscribe a cone during rotation of the blades around the rotor shaft. The cone is the result of flapping motion of the blade accompanied by rotation around the rotor shaft. Each blade flaps under competition of moments due to the blade lift forces and centrifugal forces, which appear at the blade rotation. The total lift force of all blades is almost aligned to the axis of the blade cone. The cone axis may tilt in a range up to ten degrees with respect to the rotor axis caused by the cyclic pitch as well as by azimuthally inhomogeneous blade blowing conditions. Dependence of tilt of the cone axis together with the total lift force tilt on cyclic pitch enables the control over the rotorcraft.

Blade cone tilt is a response of blade flapping motion on some cyclic impact on a blade. A cyclic impact is represented by some forces acting on a blade differently in different blade azimuthal positions. If some cyclic force acts downward on a blade in certain azimuthal position ψ_f and accordingly upward in $\pi + \psi_f$, then the blade cone responds by tilt toward azimuthal direction which is ahead on lead angle $\Delta\psi$ from ψ_f in the rotation direction: $\psi_f + \Delta\psi$. The lead angle is positive and not greater than $\pi/2$; the lead angle equals $\pi/2$ at hovering with no blade compensation and with negligibe small flapping hinge offset; the flapping compensation decreases the lead angle on $\arctan k$ at hovering ($\Delta\psi = \pi/2 - \arctan k$); flapping hinge offset as well decreases the lead angle. Such cone tilt with lead angle as a response on some cyclic impact defines the blade cone cross-coupling response here. The lead angle depends on forward velocity and does not depend on blade non-uniform parameters.

The blade cone responds to the cyclic pitch of the swashplate. The blade cone tilts not in direction of the cyclic pitch minimum; the cone tilts toward azimuthal direction with the lead angle (phase lag) from the minimal cyclic pitch azimuthal direction due to the cross-coupling response. The shift is usually eliminated by phase lag of the control links of the stationary swashplate (reverse to the lead angle). The magnitude of the cone tilt is smaller than the cyclic pitch depending on the compensation coefficient and flapping hinge offset.

The blade cone responds to rotor system rotation by a damp response and by a transverse response. The damp response is represented by blade cone tilt in opposite direction to the rotor system rotation (at no compensation and negligible hinge offset) which is caused by Coriolis forces due to mutual rotation of the rotor system with the rotor shaft rotation. The damp response is proportional to the angular velocity of this rotation and depends on the mass properties of the rotor blades: the heavier are the blades, the stronger is the response. The transverse response is represented by blade cone tilt in the direction opposite to the vector of the rotor system angular velocity (at no compensation and negligible hinge offset) which is caused by cyclic changing of element attack angles due to the circumferential velocity of this rotation. The flapping compensation and the flapping hinge offset azimuthally shift the responses in the opposite direction to the rotor shaft rotation.

The blade cone responds to the rotor system accelerated rotation by the cone tilt with azimuthal delay $\pi/2$ in the opposite direction of the shaft rotation (at no compensation and negligible hinge offset) from azimuthal direction with maximal blade accelerated circumferential motion downward due to this rotation. The response is caused by the inertial forces due to the accelerated rotation. The response depends on the mass properties of the blades: the response is stronger for heavier blades. The flapping compensation and the flapping hinge offset azimuthally shift the responses in the opposite direction of the rotor shaft rotation.

The blade cone responds to forward motion of the rotor system by backward tilt in zero azimuthal direction and by right tilt in azimuthal direction $\pi/2$. The backward tilt is caused by the cone cross-coupling response on different air velocities of advancing and retreating blades. The right tilt is caused by the cone cross-coupling response on different air blowing of front and back blades at non-zero cone angle ($a_0 \neq 0$): the front blade is blown from below by forward motion airflow, the back blade is blown from above. The cone response on forward motion increases with the speed of the motion.

9.2.2   The total rotor force is, first of all, represented by the total lift force, which accumulates lift forces created by all rotor blades and enables to carry the rotorcraft in the airspace. The total lift force is generally directed close to the axis of the blade cone, which is close to the rotor axis; therefore, the main component, which is directed along the rotor axis and is defined as the rotor thrust, represents the total lift force in terms of ability to carry the rotorcraft.

The components, which represent the projections of the total rotor force on the rotor plane and are defined as the longitudinal rotor force and the side rotor force, mainly are represented by projections of the total lift force on the rotor plane caused by the blade cone tilt. The longitudinal and side rotor forces are much weaker than the rotor thrust; however, these forces create moments about the rotorcraft center of mass, which can change orientation of the rotorcraft in the airspace. Since the blade cone tilt depends on swashplate tilt, then these moments created by the longitudinal and side rotor force enable control over the rotorcraft via the swashplate mechanism. Such control is direct at hovering and vertical motion: the swashplate tilt causes tilt of the blade cone axis; the total rotor force is tilted together with the cone axis and creates correspondent longitudinal and side rotor force, which create the moments about the rotorcraft center of mass.

Blowing conditions of blades become different at different blade azimuthal positions at forward motion of the rotor system, which contributes to the longitudinal and side rotor forces. These contributions are caused by azimuthally inhomogeneous induced drag of blade elements, which appears at blade rotation at forward motion. The advancing blade is blown by stronger airflow, and retreating blades is blown by weaker airflow at forward motion. The azimuthally inhomogeneous blowing is as well caused by blowing the blades forming the cone with non-zero cone angle in the airflow of the forward motion: the front blade is blown from below, and the back blade is blown from above. The induced velocity becomes azimuthally inhomogeneous at forward motion, which as well contributes to change of the longitudinal and side rotor forces at forward motion. The total profile drag of all elements of all blades contributes only to the longitudinal rotor force at forward motion, is proportional to forward motion speed, and is directed opposite to the forward motion.

The rotor force responds to the rotor system rotation by the damp response and by the transverse response. The damp response is represented by the superposition of the longitudinal and side rotor forces, which is directed opposite to the rotor system rotation, is proportional to the angular velocity of this rotation, and is caused by inhomogeneous blade lift forces due to this rotation at non-zero blade cone angle; the larger the cone angle, the stronger radial projections of the inhomogeneous blade lift forces on the rotor plane. The transverse response is represented by the superposition of longitudinal and side forces, which is directed along the vector of the angular velocity (laying in the rotor plane) transversely to the rotation, is proportional to the angular velocity, and is caused by inhomogeneous induced drag due to this rotation.

9.2.3   The rotor blades affect the flapping hinges, to which they are attached, by applying the blade forces on the hinges. These forces create the moment about the hub center as the reference point due to the non-zero flapping hinge offset. This hub moment consists of the lateral hub moment and the longitudinal hub moment. This hub moment, first of all, responds to blade cone tilt and is directed toward the cone tilt proportionally to the magnitude of the cone tilt. Such hub moment is caused by accelerated flapping motion of the blades forming the tilted blade cone and depends on the mass properties of the blades and the shaft rotation speed. Since the blade cone tilt depends on swashplate tilt, the hub moment participates in control over the rotorcraft together with the longitudinal and side rotor forces.

The hub moment responds to the rotor system rotation by damp and transverse responses. The damp response has aerodynamic nature and is caused by element attack angle change due to circumferential velocity at this rotation; the moment of the damp response acts in opposite direction to this rotation. The transverse response is caused by Coriolis forces due to the rotor system rotation accompanied by the rotation around the rotor shaft; the moment of the transverse response acts in azimuthal direction, which is $\pi/2$ ahead relative to the rotor system rotation, i.e. transversely to this rotation.

The hub moment responds to the rotor system accelerated rotation in direction opposite to the rotation acceleration: thus the moment damps this accelerated rotation. This moment is caused by inertial forces acting on the blades at the circumferential accelerated motion due to this accelerated rotation.

Blowing conditions of the blades at different azimuthal positions are different at forward motion of the rotor system; this causes azimuthal inhomogeneity of blade lift forces; this leads to the azimuthally inhomogeneous forces on the flapping hinges; these contribute to the hub moment at forward motion. This contribution increases with speed of the forward motion. However, the inertial components of the hub moment are dominant.

9.2.4   The rotor blades create a moment around the rotor axis during rotation together with the rotor shaft, which is directed opposite to the shaft rotation and is caused by the profile and induced drag of the blades. This moment resists the rotation of the blades around the shaft. This resistance moment affects the rotor hub together with the rotor shaft around the rotor axis but does not explicitly act on the rotorcraft.

An engine torque is needed to be applied to the shaft to compensate for the resistance moment in order to keep the permanent shaft rotation. The engine torque must equal by magnitude to the resistance moment but act in the opposite direction. The rotor shaft reacts by an opposite reactive moment on the engine, which is called the rotor torque. The rotor torque acts on the rotorcraft through the engine body which is solidly fixed to the rotorcraft. In order to keep the rotor shaft rotation constant the rotor torque equals the resistance moment on the hub and acts in the same direction.

The resistance moment represents the consumption power of the rotor system, which must be provided by the engine power in order to keep the shaft rotation constant at a rotorcraft flight. The engine power is spent on the rotor thrust generation, overcoming the blade profile drag at rotation around the shaft, to counteract the rotorcraft aerodynamic drag at forward motion, the rotation of the rotorcraft together with the rotor system. The induced power, which is spent on the thrust generation, is the dominant part of the consumption power.

9.2.5   The flapping hinge offset influences the properties of the rotor system by different arms of flapping rotation and rotation around the rotor shaft and by different moments of blade inertia around the rotor shaft and the flapping hinge. The first reason is represented by the blade-hinge characteristic parameters $B_m^{(n)}$; the second reason is represented by the ϵ_I parameter.

The flapping hinge offset decreases the lead angle of the blade cone cross-coupling response as was shown in 5.4.1. The hinge offset slightly changes the magnitude of the blade cone response; the change depends on correlation of the inertial blade properties affected by the offset and the arms differences due to the offset.

The impact of the flapping hinge offset on the total rotor force is caused by no creation of lift force in the area between the hub center and blade flapping hinges, as well as by this area does not participate in the flapping motion of the blades. This area of the rotor disc does not participate in creation of the longitudinal and lateral forces. The contribution of the flapping hinge offset in the total rotor force is insignificant and mostly depends on the blade cone tilt.

The flapping hinge offset is the main reason for the lateral and longitudinal hub moments: the hinge offset is the arm for the blade forces on the hinges, which create these moments.

9.3   REMARKS

9.3.1   The blade element rotor theory is not able to determine the induced airflow anywhere on the rotor disc. The theory provides the calculation of the dynamic properties of the rotor system with the known finite set of inflow ratio characteristic parameters, which describe the induced airflow velocity over the whole rotor disc. This is the main reason that the blade element rotor theory is not self-sufficient to determine the dynamic properties of the rotor system by itself.

9.3.2   The presented calculations of the rotor system properties are only related to a steady rotation of the blades around the rotor shaft. Such ideal conditions appear quite seldom during a real helicopter flight. A helicopter simulator must reproduce transient processes, control system time delay, as well as change of forward motion direction, which are very important for simulator purposes. These effects were skipped here; however, the presented calculations are the appropriate background for a model of a main helicopter rotor for a simulator software.

References

1. Barnes W. McCormick, Jr., Aerodynamics of V/STOL Flight. Dover Publications, Inc., Mineola, New York, 1999.

2. A.R.S. Bramwell, G. Done, and D. Balmford. Bramwell's helicopter dynamics. 2nd ed., Butterworth-Heinemann, Oxford, 2001.

3. W. Johnson. Helicopter Theory. Princeton, New Jersey: Princeton University Press, 1980.

4. M.L. Mill, A.V. Nekrasov, A.S. Braverman, L.N. Grodko, and M.A. Laykand. Helicopters. Calculation and Design. Book 1, Aerodynamics. Mashinostroyenie, Moscow, 1966.

5. Volodko A.M., Verkhozin M.P., Gorshkov V.A. Helicopters, 1992.

6. Romasevich V.F., Samoilov G.A. Practical aerodynamics of helicopters, 1982.

7. Esaulov S.Yu., Bakhov O.P., Dmitriev I.S. Helicopter as a control object, 1977.

Index

Printed and bound by CPI Group (UK) Ltd, Croydon, CR0 4YY

24/10/2024

01778605-0001